D1084404

WHEN CULTURE
AND BIOLOGY COLLIDE

WHEN CULTURE AND BIOLOGY COLLIDE

Why We Are Stressed, Depressed, and Self-Obsessed

E. O. SMITH

Rutgers University Press
New Brunswick, New Jersey, and London

Library of Congress Cataloging-in-Publication Data
Smith, Euclid O.
    When culture and biology collide : why we are stressed,
    depressed, and self-obsessed / E.O. Smith.
        p.   cm.
    Includes biblographical references and index.
    ISBN 0-8135-3103-9 (cloth : alk. paper)
    1. Human biology—Social aspects.   2. Human evolution.
    3. Social changes.   I. Title.
    GN298 .S55 2002
    304.2—dc21

                                                    2001058682

British Cataloging-in-Publication information is available from the
British Library.

Manufactured in the United States of America

To Cindy

and to the memory of my mother,
May Ethel Therrell Smith

Contents

Acknowledgments

There are a number of people who helped with this book, but the most important was my wife, Cindy Gelb. She made the prose better and the ideas clearer. She also was incredibly understanding through the final stages of the preparation of the manuscript when I was tired and grumpy.

A number of my colleagues have helped both directly and indirectly on this project. Peter Brown recognized that this book would not happen without time away from teaching. I have profited from numerous discussions with my colleagues, George Armelagos, Peter Brown, Mel Konner, Michelle Lampl, Dan Sellen, Pat Whitten, and Carol Worthman. In addition, I thank Daniel Lende, Melissa Melby, Ryan Brown, and Dan Hrushka for their patience in listening to many of the ideas and helping me refine the arguments. Several undergraduate students have also helped, doing some of the footwork that is a necessary part of this kind of a project— E. A. Quinn, C. Quave, and S. Bew paid attention to the details when it counted.

I want to thank the staff of the Robert W. Woodruff Library at Emory for their invaluable assistance in collecting the obscure and sometimes baffling references and questions that went with them. I am especially grateful to Greta Boers, Reference Librarian for Anthropology, who was always cheerful and eager to help with even the most bizarre questions.

Finally, Woody, Ivy, and Farrah saved the day. They were always glad to see me, even if all else went to hell.

Atlanta, Georgia
24 August 2001

WHEN CULTURE
AND BIOLOGY COLLIDE

1 | Intersection of Biology and Culture

> To understand culture one must have some
> understanding of biological thought.
> —William S. Beck, *Modern Science and the Nature of Life*, 1957

What do road rage, tattooing, cosmetic surgery, depression, addiction, obesity, and aid to strangers have in common? It is tempting to say nothing, but I hope to convince you by the time you have finished this book that a common thread unites these disparate topics. In addition, I hope you will appreciate that the same underlying principle is likely to have much wider usefulness in our attempt to understand the basic elements of human behavior. What these complicated sets of behaviors have in common is that they are examples of how our evolved biology (our genetic programming) interacts with our cultural environment.

There is no denying that our genes guide some of our behavior, but the question is, how much? And then, how can our culture influence what our biology urges us to do? And can our biology and culture influence each other in reciprocal ways? These are all-important questions that help us understand why we do what we do.

To appreciate more fully the interrelationship between our culture and our biology, let me offer a real-world example. The heroic acts of emergency personnel as well as countless numbers of civilians following the terrorist attacks on New York City and Washington, D.C., have been recounted on television and radio, in newspapers and magazines, around watercoolers and coffee machines, and at barbershops and beauty salons throughout the United States. For example, many Americans have heard about thirty-six-year-old Michael Benfante, a New York branch manager for a telecommunications firm, who worked on the eighty-first floor of 1 World Trade Center. On the morning of 11 September 2001, he and a colleague got a woman into an emergency wheelchair and carried her sixty-eight stories down the stairs through smoke, fire, and falling debris to a waiting ambulance.

As the tragedy unfolded at Ground Zero, Todd Beamer and several other passengers on United Airlines flight 93 overpowered the hijackers that had taken control of their aircraft and, by their heroic actions, likely saved hundreds or more lives. Beamer will be remembered for his "let's roll" call to arms to several other men on the flight, which then crashed in rural Pennsylvania.

These self-sacrificing acts are the stuff of which legends, as well as movies, are made. That we have the ability to overcome our innate selfish tendencies is a defining characteristic of humanity. Not only do we have the capacity to treat other humans in a self-sacrificing manner; we exhibit altruism to nonhumans as well. Immanuel Kant once said, "We can judge the heart of a man by his treatment of animals." If Kant is right, then there are some three hundred or so hugely altruistic owners/handlers of the search-and-rescue dogs that worked tirelessly at Ground Zero. The unique bond between dog and master has been shown in photographs of exhausted German shepherds curled up asleep next to their handlers in emergency shelters near the collapsed towers. Dogs and their handlers worked twelve-hour shifts in an attempt to find victims of the attack buried under the tons of steel and concrete. Not only were there countless numbers of volunteers aiding the emergency personnel at Ground Zero; there were volunteer veterinarians who set up emergency treatment facilities for the search-and-rescue dogs. Like the handlers and the dogs, the vets also worked twelve-hour shifts, treating cut paws, dehydration, and respiratory distress. The American Humane Society collected thousands of dollars to help not only the search-and-rescue dogs but also the pets displaced by destruction.

The outpouring of selflessness in the wake of the tragedies in New York, Washington, D.C., and Pennsylvania reminds us of this capacity within each of us. In stark contrast are the acts of the terrorists, who gave their lives in the service of a religious fanatic. Their self-sacrifice, unlike the altruism of those giving aid and assistance, was profoundly selfish. Focused as they were on the destruction of the enemy and their ultimate reward of an eternal afterlife in Paradise (al-Jannah), the terrorists demonstrated the limits to which culture can push our basic biology. In a real sense, the attacks on the World Trade Center and the Pentagon show us how powerful culture can be in reinforcing certain basic human predispositions, such as selfishness, as well as how powerful culture can be in overcoming those same predispositions.

Understanding human behavior is a difficult but not an impossible task. Unlike other animals, humans rely on the transmission of information across generations. This is not to say that other animals and birds do not communicate with one another, for there is a considerable body of scientific literature to the contrary. Yet we humans rely far more on the intergenerational transmission of knowledge. The advantage that this gives us is extraordinary. We do not have to reinvent skills, techniques, and beliefs anew each generation. The cumulative effect of this fundamental change in human culture is unparalleled, to the best of our knowledge, in any other animal. While other animals, particularly chimpanzees, transmit specific information from parent to offspring, and offspring learn by observing and practicing parental behavior, no other animal relies on this kind of information transmission as much as humans. It is the centerpiece of our culture.

It seems to me that understanding human behavior without understanding something about this information transfer is impossible. First, we must define the body of information that we pass from parents to offspring and among members of the same generation, age cohort, clan, lineage, moiety, and so on. Anthropologists have worried a lot about how to describe culture, but a classic definition written over a century ago still provides an excellent starting point for discussion. Edward Tylor, a British social anthropologist, wrote in his book *Primitive Culture* (1871): "Culture . . . is that complex whole which includes knowledge, belief, art, morals, law, custom, and any other capabilities and habits acquired by man as a member of society."[1] Now the important part of this definition is the last part. What I am talking about are the cultural practices that we acquire as members of society.

Our behavior is a product of our culture and our biology. We live our lives immersed in a culture, and in some ways culture transforms us, but we also transform our culture. What we bring from biology to the cultural arena is a set of basic predispositions to behave in certain ways. The outcome of this interaction is the subject of this book.

The life-and-death choices that parents make about the health and well-being of their children are a classic example of the intersection of biology and culture. These, too, can partake of the tragic.

On 7 July 1989, for instance, a Pakistani woman brought her infant twins into a clinic in Islamabad. She had given birth to her fraternal twins five months earlier. At birth, her mother-in-law took the female twin away and bottle-fed her. The mother kept and breast-fed the male twin. The daughter suffered from malnutrition, dehydration, diarrhea, and respiratory problems, while the son was well nourished and in apparent good health.

In a matter of hours after arriving at the clinic, the female twin died; malnutrition and dysentery had consigned her to death at the hands of the mother-in-law.

Understand that the mother-in-law was not a deranged child abuser. Instead, she was acting out a behavioral program that from an evolutionary point of view was perfectly sensible. Helping rear her grandchild was natural for her since a mother of twins faces an exceedingly difficult time in providing adequate nutrition for a single infant, much less twins. The "natural causes" of the death are tragic to us, but to many women around the world this kind of benign neglect is commonplace. In fact, many use much more drastic measures to get rid of unwanted sons or daughters.[2] Don't get me wrong. I am not advocating infanticide or neglect of infants, but both are unpleasant facts for many people around the world. My goal, rather, is to explore how culture could lead a mother, whose biology wants only the best for her babies, to the point of giving up one of her offspring to what she must know is a likely death.

"Tis a happy thing to be the father to many sons," William Shakespeare wrote in his play *King Henry VI*.[3] The preference of parents for sons over daughters has a long history among humans and is a tradition that spreads across many cultures worldwide. A Tibetan proverb noted that "daughters are no better than crows; parents feed them and when they get their wings they fly away." Nevertheless, why are sons valued so highly? This is a hot topic in the world of evolutionary anthropology, and there is no firm agreement among those who study human parenting. Mead Caine, then a researcher of the Population Council of New York, quantified the value of labor provided by sons as compared with daughters in Bangladesh in a study published in 1971. He showed that the value of labor contributed by sons to parents greatly exceeds that of daughters. By age fifteen a son has repaid his parents their investment in his rearing; by age twenty-one he has also paid for a sister. While daughters work hard and contribute to the family well-being, they leave their natal family too early to repay parental investment.[4]

So, how does this relate to the intersection of culture and biology? Over the history of the human species, and also among other animals and birds, parents have decided the allocation of resources to offspring. Almost thirty years ago, Robert Trivers, then a young professor at Harvard, developed the idea of parental investment. That is, parents who provide more resources to their offspring enhance the likelihood that their offspring will survive, all else being equal. Now, according to Trivers, the resources that parents can generally provide cannot be shared among offspring. Food given to one, for example, cannot be given to another, although protection could possibly be a kind of shared parental investment.[5]

Clearly, human parenting involves deciding investment in offspring. Decisions to have children or not, how many to have, and how to allocate resources among existing offspring, as compared with the production of future offspring, are all decisions that humans engage in every day. With the Pakistani woman and her twins, it is unlikely that the mother could have mobilized the necessary resources to feed both of her children adequately. Given the strong cultural preference for sons in Pakistani culture, the mother made a clear decision to allocate her care preferentially. This may seem like a cruel and heartless way to treat a child, but, given severely limited resources, saving the child that is likely to give greater return to the parents is more advantageous, in terms of both short-term economics and long-term genetic survival (representation of their genes in future generations).

There are those in modern biology who are not concerned about such evolutionary explanations for our behavior, explanations that take into account both our biology and our culture. They seek only those explanations firmly grounded in molecular biology. These reductionist explanations have provided important insights into much of human anatomy and physiology, and to some extent our behavior, but they ignore the larger *why* questions. Admittedly, the *how* questions, although not easily solved, are much more easily attacked using the modern scientific method. Observation, manipulation, and replication are the hallmarks of the experimental method and, on the molecular level, have yielded truly amazing insights into the human condition. However, as the Nobel prizewinning zoologist Niko Tinbergen once wrote, to understand completely any biological phenomenon, one must understand its evolution, its function, its underlying (proximate) cause, and its ontogeny (development in the individual).[6] Therefore, since there is not just one explanation for any phenomenon of interest, to think that a single explanation of any human behavior is adequate is shortsighted at best. Several explanations may all be correct simultaneously but have relevance at different levels.

To fulfill Tinbergen's admonition, I want to start our journey toward a better understanding of human behavior several million years ago. We know that more than four million years ago, in East Africa, early human ancestors evolved an upright posture.[7] The suite of anatomical changes that allows us to walk bipedally ranks with some of the most radical adaptations in human history. Bipedalism involves complicated and extensive changes to the internal organs, the pelvis, the arms, the lower legs, and the spinal column. These changes are unparalleled in other primates or other mammals and their likes can be found only in a few birds.[8] Interestingly, changes in locomotor patterns preceded other profound changes (e.g., increase in brain size, long gestation, etc.) separating us from other primates. Besides bipedalism, our lineage underwent strong selection for increased cognitive abilities. We know this because of the evidence that paleontologists have discovered dating back to the Pliocene. It is now well recognized that members of the lineage that likely led to modern humans showed a dramatic increase in body size and brain size about two million years ago. Coincident with this fundamental change in brain size and body size was the manufacture of the first stone tools, well known archaeologically from several sites in East Africa that date as early as 2.6 million years old.

What seems clearer and clearer with each passing field season among human paleontologists is that our ancestry was far from unilinear but was very much as Charles Darwin conceived it almost 150 years ago. Darwin used the analogy of a branching bush to describe the process of evolution. The more information that becomes available from excavations in fossil-rich formations in Kenya, Tanzania, Ethiopia, Georgia (old Soviet Union), China, and the Middle East, the more confusing the evolutionary story becomes.

It is with good reason that human paleontologists create much ado about a single fossil molar, a distal portion of a right humerus, or a fragmentary occipital. When the probability of finding a human fossil of great antiquity, in a condition making examination and study even remotely possible, is vanishingly small at best, wringing out all the data possible from each specimen is important. The net result of these efforts has been the proliferation of human family trees, and today there are almost as many different phylogenies as there are human paleontologists. The important point is not whether one evolutionary tree is right and the others wrong; it is to recognize the amazing amount of variation that existed among our early ancestors. What we likely are seeing in the human fossil record are many evolutionary experiments that failed and only a few that succeeded. Precise identification of the winners and losers is not yet possible.

It is important to remember that the biological species *Homo sapiens* is the result of 180 million years of mammalian evolution, 65 million years of primate evolution, 5 million years of hominid evolution, 2 or so million years of the evolution of our genus, and no more than 150,000 years as *Homo sapiens*. The exact timing and details surrounding the evolution of the first humans are contested and are well beyond the scope of this book. The important thing to note is that over the last 150,000 or so years, the development of tool technologies and the subsequent changes in the human environment have moved with a rapidity not previously experienced in the entire evolutionary course of humans.[9]

Since the first settled villages dependent on agriculture appeared only about 10,000 years ago, the vast majority of our evolutionary history took place in the context of a nomadic lifestyle, with hunting of wild game and gathering of vegetable foods.

Archaeological evidence suggests that we functioned in small groups and with a simple technology until very recently. Survival of the fittest in that archaic environment profoundly shaped our biology—our dietary, health, emotional, and also our psychological needs.

With the onset of agriculture, dietary diversity and health actually decreased significantly.[10] Instead of the wide array of wild plant foods that were the staples of foraging societies, a narrow range of cultivated crops made up as much as 90 percent of diets, with considerable negative consequences for health. Wild plant foods were the mainstays of preagricultural people except for inhabitants of the highest latitudes. Roots, beans, nuts, tubers, fruits, and gums gave the hunters and gatherers a vastly greater variety of vitamins and minerals that those available from cultivated cereal crops.

Preagricultural people led lives that required much more physical exertion than does our modern Western lifestyle. A reasonable estimate of their average daily energy requirement was about three thousand calories. Their vegetable intake would have been bulky, diverse, and nutritious. The meat they consumed from wild game animals was 4 percent fat compared with 29 percent fat in corn-fed livestock, and meat of any type was much less abundant than it is today.[11]

Our evolutionary heritage is all around us, and it is particularly conspicuous in our dietary choices. In our evolutionary past, our ancestors ate roughly half the fat that we do but three times the protein. The fat that they consumed was more polyunsaturated than saturated. They had few refined carbohydrates and no finely ground flour in their diet, and, if they ate grains, they ate whole grains. Their sodium intake was one-fourth of ours, and they consumed more potassium than sodium, the opposite of our dietary pattern today. Their calcium uptake was twice as great as ours. Whole grains provided an abundance of essential micronutrients, and such a diet was likely bulky and more filling than one containing modern, refined forms of food.

Yet agriculture, despite its nutritional drawbacks, allowed more people to live on less land. In preagricultural days, the population of the earth doubled about every fifteen thousand or so years. With the rise of agriculture, the population of the earth doubled every thousand years (see fig. 1-1). Now the population doubles about every thirty-five years. In 1993, the population of the earth was approximately 5.5 billion, then in 2000 6.15 billion. It will be an estimated 8.5 billion in 2025 and an estimated 9.1 billion by 2050.[12]

The story of the Pima Indians, and especially their changing diet, perfectly illustrates the conflict between biology and culture. Native to southern Arizona in the Salt and Gila River valleys, Pima Indians are known for their manufacture of extraordinary basketry as well as their sophisticated agricultural practices.[13] Unfortunately, they are also known for high rates of diabetes, a sad intersection of biology and culture. To explain what happened, I have to diverge a bit and discuss diabetes.

Insulin and glucagon are the main hormones secreted by the pancreas, and in combination they control blood levels of glucose (see fig. 1-2).[14] Glucose, the basic blood sugar molecule, must be maintained at a constant level in the blood (80–120 mg of glucose/100 ml blood) for normal, homeostatic body functions. When glucose levels are high, right after a meal, the pancreas releases insulin, which lowers blood glucose levels by stimulating

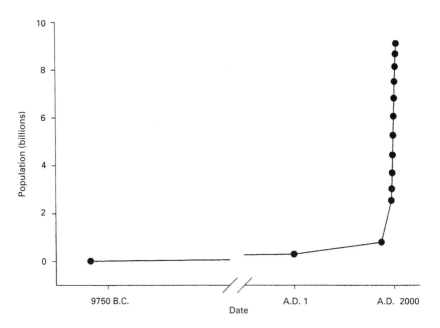

Fig. 1-1. World population growth from roughly the origin of agriculture to present day. © E. O. Smith, MMI.

skeletal muscle and other cells to absorb glucose, thereby removing it from the bloodstream and rapidly converting it to glycogen, a glucose storage molecule. Insulin also stimulates fat cells to absorb glycogen and convert it into fat. On the other hand, when glucose levels are low, due to lack of food or to increased exercise, there is a demand for increased glucose levels to fuel the body. The pancreas secretes glucagon, which stimulates the liver to

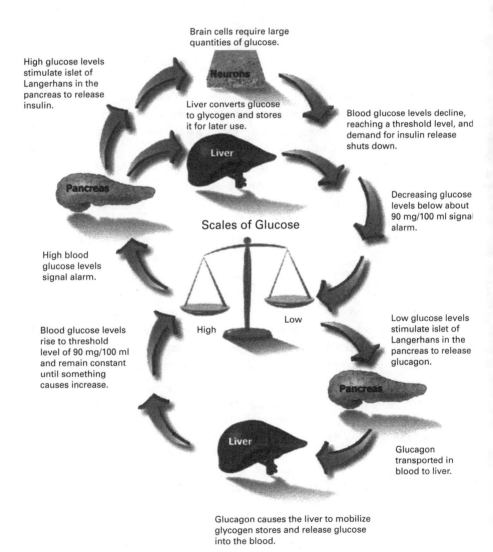

Fig. 1-2. Glucose regulation is an example of a negative feedback system. Diagram adapted from Marieb 1995. © E. O. Smith, MMI.

convert glycogen back to glucose and release it into the blood. Glucagon also stimulates the liver to convert amino acids, glycerol, and lactic acid into glucose, thus providing adequate energy supplies to all the vital organs.

Blood sugar needs to be maintained at ninety milligrams per hundred milliliters. How does the body regulate blood sugar levels so precisely? The answer lies deep within one of the most versatile organs in the body, the pancreas. The pancreas supplies digestive enzymes to the small intestine, but it is also responsible for monitoring blood sugar levels. Inside the pancreas, there are clumps of cells called the islets of Langerhans that do not release their secretions directly into the pancreatic duct but secrete hormones instead. The islets of Langerhans are composed of two different cell types: alpha cells, which secrete the hormone glucagon, and beta cells, which secrete insulin. When blood sugar levels drop below the threshold value of ninety milligrams per hundred milliliters, the alpha cells are stimulated to secrete glucagon, raising blood sugar levels back to the threshold value. When blood sugar levels reach the threshold value, the alpha cells shut down. If blood sugar exceeds ninety milligrams per hundred milliliters, then the beta cells are stimulated to secrete insulin, reducing blood sugar levels. Therefore, a negative feedback mechanism maintains blood sugar levels.

If, however, blood sugar remains too low (hypoglycemia) or too high (hyperglycemia), there are problems. Hypoglycemia results in organs being starved of energy. The brain, a massive consumer of glucose, is especially vulnerable. When the brain is starved of glucose, hypoglycemia can be fatal. Symptoms include weakness, tremors, convulsions, disorientation, and unconsciousness. On the other hand, hyperglycemia results in diabetes, which is characterized by high glucose levels in urine (as high as 1,200 mg/100 ml) and production of massive quantities of urine accompanied by severe thirst. Diabetes results from inadequate production of insulin by the beta cells in the islets of Langerhans. Defective beta cells can be the result of childhood viral infections such as mumps or measles. Genetic inheritance may also play a role, in that some individuals are born without functional beta cells. When this happens, the beta cells cannot secrete adequate amounts of insulin, and the body cannot absorb glucose (and hence cannot store it as glycogen for energy). The body responds by drawing on fat reserves for energy, and that results in the production and release of acidic ketone bodies. These ketone bodies cause the breath of a diabetic to smell like acetone. The ketone bodies cause the blood acid level to rise (acidosis), which can cause unconsciousness (a diabetic coma) and even death.[15]

Now imagine that there existed a mutation that promoted the production of insulin by beta cells but without the conversion of glucose to glycogen and its storage for later use. This means that the body would be "resistant" to insulin. Glucose levels in the blood would rise as usual with ingestion of food high in carbohydrates, but glucose levels would then remain high far longer. Individuals with insulin resistance would have a potential metabolic advantage over those less able to maintain normal circulating glucose levels. During periods of low food availability or starvation, individuals with the mutation could maintain normal glucose levels even in the face of a reduced caloric intake.[16]

Pima Indians have such a genetic makeup, which permits insulin to remain in the blood without lowering glucose levels. James V. Neel, a human geneticist at the University of Michigan, made this extraordinary discovery while investigating the genetics of "isolated" human populations. Neel called this condition the "thrifty genotype." When isolated peoples were faced with starvation conditions, natural selection would have highly favored this mutation. In fact, the Pima are not only resistant to insulin's ability to promote glucose storage in the liver and muscles; natural selection has also built this resistance up as a preference for fat stores instead.

Now this "thrifty genotype" would have been highly adaptive in the evolutionary past, when it is likely that the ancestral Pima population experienced a long period of reduced food availability and starvation. A few individuals possessed this curious mutation, which gave them an enormous advantage in the face of food scarcity. The ancestral population must have hit what geneticists call a bottleneck,[17] when most of the individuals without the mutation died. The surviving individuals essentially established a new genotype for the Pima that included many individuals with the "thrifty genotype."

Genes that were advantageous to Pima Indians in the past, however, are the same ones that predispose them to diabetes today. High levels of glucose are easily maintained in the bloodstream, and those high levels can lead to diabetes. About one-half the Pima Indian population has diabetes, and, of the diabetics, 95 percent are obese. To understand what has gone on with the Pima Indians, knowing a bit about their culture is important.

Traditionally, the Pima lived in settlements near the riverbanks of the Gila and Salt Rivers and practiced intensive agriculture characterized by the construction of irrigation canals. The Pima Indians maintained their traditional farming lifestyle raising maize, beans, and squash, along with some livestock, until the late nineteenth century. Fishing and small game hunting also supplemented the traditional diet, along with gathering wild beans and fruits. Then, American farmers settling upstream began diverting the

Gila River water supply. The Pima Indians' two-thousand-year-old tradition of irrigation and agriculture was disrupted, causing poverty, malnutrition, and even starvation. The Pima community had to rely instead on lard, sugar, and white flour doled out by the United States government. However, World War II brought great social and economic change for American Indians overall and the Pima in particular. Many entered military service.[18] Others left the reservations to take jobs in the war industries that had developed in California. The economic gains were dramatic for the Pima Indians, with estimated cash income more than doubling between 1940 and 1944. Along with the diet of many Americans after World War II, the diet of Pima Indians changed dramatically. In the 1890s, the traditional Pima Indian diet consisted of only about 15 percent fat and was high in starch and fiber; currently almost 40 percent of the calories in the Pima diet are derived from fat. As the typical American diet became more available on the reservation after the war, people became more overweight.

Several factors contribute to the prevalence of diabetes among the Pima Indians today including diet, inactivity, obesity, and alcoholism. The relationship between alcohol consumption and diabetes is through the damage that alcohol does to the pancreas. The precise way that alcohol induces tissue necrosis is unclear, but the damage results in the inability of the pancreas to secrete insulin and glucagon. The disruption of normal pancreatic function coupled with a genetic predisposition to elevated serum glucose levels leads to serious health complications for many Pima Indians.

Recently, studies of a Pima community practicing a traditional lifestyle, living in a remote area of the Sierra Madre Mountains of Mexico, offered some interesting comparisons to the Pima in Arizona. The Mexican Pimas are genetically identical to the Pima Indians of Arizona. However, instead of the expected prevalence of diabetes, 34 percent among females and 57 percent among males, researchers found only three of the thirty-five Mexican Pimas studied had diabetes (slightly more than 8 percent), and the population as a whole was not overweight. This suggests that lifestyle modification among the Arizona Pimas may offer significant relief from diabetes.[19]

I am certain that this discussion about the Pima Indians has seemed like a lengthy diversion into gastrointestinal physiology and pathology, but it serves to illustrate the idea of "thrifty genes." The underlying idea of a specific modern lifestyle conflicting with one particular genotype is deceptively simple, yet it has significant power in helping us understand the more general conflict between our culture and our biology. The profound changes that have occurred in the cultural environment of our species since we arose a few thousand generations ago are of considerable interest. The last few thousand years of human history have witnessed dramatic changes in our mobility, eating habits, modes of production, patterns of communication—our fundamental way of living. However, there is little evidence to suggest that over the last few thousand generations we have changed our fundamental biology in any significant ways. To be sure, there are examples of fine-tuning between our biology and our environment (such as adaptations to malaria in Africa and Asia and to lactose intolerance in northern Europe) but the basic plan remains the same. The rate of biological evolution approximates the speed of a glacier in comparison to the rate of change in our culture. This vast discrepancy in rates of change sets the stage for the paradoxical ways that biology and culture interact.

We live in a cultural environment that is in many ways a response to our Pleistocene genetic heritage. A common complaint among a small minority of students in my classes the first day of a new semester is the lack of left-handed desks in the lecture halls at my university. We live in a right-handed world.[20] Now the interesting question is why so many of us are right-handed. An evolutionary perspective would suggest that there must have been something in our evolutionary past that favored righties over lefties, but what could it have been? The answer to that question is beyond the scope of this book, but it does illustrate the power of evolutionary thinking.

Another common question on the first day of class concerns my attendance policy. I make all my lectures available to students electronically, so theoretically, according to student logic, they don't necessarily have to come to class. I try to emphasize that, in the past, students who regularly attended class did better on the exams than those who cut. Why should that be the case? There are commonsense answers that, I would argue, are based in our biology. While it is true that we are designed to rely heavily on visual sensory information, for the vast majority of our evolutionary history there was no written language. We also have highly developed brains to process auditory information. Consequently, it should come as no surprise that students who supplement the written sources of information with audition during lectures tend to be most successful. It is also true that, even if students recognize this fundamental truth, they have to stay awake for the benefits to accrue.

Yet we remain highly visual animals. Is there any wonder that graphics are such an important part of our educational methods? Educators have long recognized this idea, and, with the technology explosion of the last decade, several sophisticated computer devices have become available to make the job of creating graphics for educational purposes much easier. Why do my students respond so positively to my use of computer-generated graphics and video in my lectures? Our Pleistocene brain is wired to deal with graphical representations in an extremely efficient manner.[21] It is likely that our perception of the world around us, our cognitive organization, and our emotional responses have been shaped to some extent for a Pleistocene environment. Given these evolutionary biases, it is easy to see how quickly we might revert to old patterns of behavior, even in a vastly different modern environment.

The central theme of this book is the intersection of biology and culture, which often seem to conflict, but the underlying questions throughout ask *why?* It has long been recognized in evolutionary biology that there are many different answers to the *why* question, and it is important to keep them straight.

Causation is a cornerstone of the scientific method. Problems of Newtonian physics have referenced questions of causation in its simplest form. Researchers can establish definite cause-and-effect relationships in physics. In biology, however, this kind of simple, straightforward relationship is rarely the case beyond the cellular-molecular level. Often our search for a single cause leads us around and around in circles, trying to decide the cause of a long and complex series of events. The fact that every phenomenon or process in biology is the outcome of two separate conditions further complicates this problem: there is the immediate, functional cause and the evolutionary cause. We call these the proximate and ultimate cause. Offering both a proximate and an ultimate cause for any observed phenomenon is usually possible.[22]

Take, for example, the modest degree of sexual dimorphism in modern humans. Females are reported to be, on average, 5 to 8 percent shorter than men in all populations, while males tend to be heavier (about 10 percent) and, on average, have a larger brain (1,300 cm³ vs. 1,400 cm³).[23] Now there might be a variety of functional causes for these metric differences between males and females that could include differences in food availability both as children as well as adults, differences in workload and energy expenditure, differences in growth hormone from the pituitary, growth-hormone releasing factor (GRF), and somatostatin from the hypothalamus. What these possible causes have in common is that they all can explain the functional differences in female and male body size, but they do not explain *why* the differences exist. It has been suggested that these differences arose long ago in our evolutionary past. The modest differences between human males and females are deeply rooted in our ancestral species' polygynous mating system and are an adaptation to a Pleistocene environment.[24] Males that were slightly larger than females may have had superior competitive abilities, and that difference was accentuated, not only by competition among males, but also by the effects of the choice of larger males by females. Now whether any of the suggested proximate or ultimate explanations are correct is the province for scientific debate, but the point should be clear that there are multiple levels of explanation with which we must be concerned if we are to understand our biology and behavior.

If we assume that there is (or was) something adaptive about the modest sexual dimorphism in humans, we are right to ask whether the degree of sexual dimorphism we see today is adaptive (based on its current function) or simply due to historical circumstances. Now many evolutionary biologists are concerned only with the current function of any observed phenomenon, not the history of changes in function. It is also true that organs/ systems/phenomena may exhibit one function that serves to maintain them in the modern population but that the function could have changed rather dramatically during the evolutionary history of the organ/system/phenomenon. Adaptation and consequently the ultimate explanation can be defined in terms of either the historical function in the organism or the current function, if the distinction is made clear.

Now to make the water a bit cloudier on this issue, we need to consider the possibility that not all observed organs/systems/phenomena may be adaptive in their current form, but nonetheless they may persist in the population. Adaptations are not perfect, and they are even less likely to be perfect on all levels of organization simultaneously. Adaptations may be imperfect due to a time lag. Environmental changes may occur, but it may very well take time for natural selection to act, so for a while the current adaptations will be out of date.

Organs/systems/phenomena may be imperfectly adapted to their environment for a variety of other reasons including genetic constraints, developmental constraints, or historical constraints. The important point is that what we see today in any observed organ/system/phenomenon is the outcome of a trade-off among the various adaptive needs of the organism.

It is precisely in this way, searching for the function of certain aspects of human behavior but mindful that nonadaptive traits/behaviors can exist, that I want to view some of the curious intersections between biology and culture in modern humans. So far our discussion has focused on the adaptive nature of traits, and how traits could persist in a population long after their fitness-enhancing attributes have changed.[25] A way that imperfect adaptations can occur or persist is explained by the "discordance hypothesis."[26]

According to this hypothesis, the rate of cultural change since the evolution of modern *Homo sapiens* is astonishing, but, in the face of that rate of change, there has not been sufficient time for many new and successful genetic adaptations to occur. Actually, even if significant adaptations could occur in such a short time, the rate of cultural change that we now experience is so rapid that even the newest and most improved adaptations would be outdated by the time they were expressed in modern populations.

It is true that some adaptations have occurred in the last ten thousand or so years (sickle-cell hemoglobin for protecting against malaria and the presence of lactase for digesting nonhuman milk in many populations), but in essence we are operating with a body adapted to what life was like thousands of years ago while we are now immersed in a culture that is radically different from generation to generation. The discordance hypothesis suggests that this is not a trivial problem, and some chapters that follow speak directly to this issue. In other chapters, I do not speak directly to the discordance hypothesis, but the careful reader can see clearly how these differences between our evolved biology and modern culture pose curious and often uncomfortable situations for us.

A number of human behaviors having to do with sexual reproduction are not easily understandable by the discordance hypothesis. As a theoretical underpinning for the discussion of such behaviors, related to dieting and the quest for thinness, body decoration, and fashion, we must look to sexual selection, a type of evolutionary change that operates side by side with natural selection.

Darwin recognized that there exist a number of phenotypic traits in the animal world that we cannot explain as the result of natural selection weeding out nonadaptive traits (e.g., tails of peacocks so heavy they inhibit flying, oversize antlers of giant elk, bright and conspicuous coloration of many fish). He reasoned that, in sexually reproducing species, each sex was constrained in its evolution by entirely different factors. He further reasoned that, rather than being linked directly to copulation, the constraints depended on the role of individuals in mating and courtship. For the *courting* sex, he noted that competition was the major driving force. Members of the courting sex competed with each other to determine who was the most vigorous, and they evolved traits to display this healthy vigor: large tails, antlers, brightly colored patches, and so forth. The most vigorous members of the courting sex were the ones that gained mating access to members of the opposite sex, the *courted* sex. On the other hand, members of the courted sex did not compete with each other (since they were the sex in short supply),[27] and, as Darwin reasoned, the members of the courted sex would select their mates from among the most successful of the courting sex. Darwin called this twofold process sexual selection.[28]

The uniting theme of this book is the role that evolution plays in the expression of modern human behavior. Our physiology and behavior are the products of thousands and thousands of generations of evolutionary history. Every day we play out behaviors that have been a part of the human experience for a very long time. Yet these behaviors are played out in an arena that is far removed from that in which they evolved. This lack of synchrony between behavior and environment is grist for the evolutionary mill. I argue that this discordance between behavior and environment sets up conditions in which there can be real conflict between our evolved psychological predispositions and the dictates of modern culture.

In addition to conflicts arising from the disjunction of biology and culture, there is another ongoing dialogue and debate between our culture and our biology over sex. We are subject to the forces of sexual selection just like all sexually reproducing species. In the larger picture, the power of sexual selection is not as strong among humans as, say, among elephant seals, but it is nonetheless quite real. So, in addition to the conflicts predicted by the discordance hypothesis, conflicts between natural selection and sexual selection are also apparent.

The case studies I have chosen to review in the following chapters were selected because of their visibility and seeming importance in our everyday lives. My goal is not to suggest an evolutionary explanation for all of our personal and societal ills but to offer a perspective that may be useful in developing intelligent solutions to what seem to be intractable problems. In addition, I am not suggesting that an evolutionary explanation should be taken as an excuse to engage in certain undesirable behaviors. To say that we are predisposed to be aggressive, so dangerous driving practices are justified, is nonsense. Understanding something about how we have evolved to express and manage our aggressive behavior may allow for the development of alternative methods to modify behavior. One of the risks that a book like this runs is that people will use it to rationalize dangerous and antisocial behavior. I hope that instead it will allow us to see ourselves more clearly and develop strategies for minimizing those destructive tendencies.

2 | Road Rage, Stress, and Evolution

> Have you ever noticed? Anybody going slower than you is an idiot, and anyone going faster than you is a maniac.
> —George Carlin

Startling headlines have appeared in newspapers across the United States reporting unbridled aggression on streets and highways. Such aggression has been the lead story on the evening news; police have charged a football hall of famer with it,[1] a heavyweight champion boxer has been convicted of it, the *Oxford Modern English Dictionary* now includes its definition,[2] and it has been the topic of a hearing before the Surface Transportation Subcommittee of the United States Congress. The subject of all of this attention is road rage.

Salt Lake City—Seventy-five-year-old J. C. King became irritated at Larry Remm, forty-one, when Remm honked at him for blocking traffic. King followed Remm and forced him to pull off the road. King threw a prescription bottle at Remm and then smashed into Remm's knees with his car, a 1992 Mercury.

Milwaukee—Shuron C. Davis, 24, was riding with a friend, Duran S. Hills, 23, when they were cut off by another car in traffic. Harsh words were exchanged which only served to increase the level of anger. The two cars jockeyed for position along a street on the north side. Davis grabbed a Russian-made assault rifle that he owned from the back seat of the car, leaned out the window and fired several shots into the other car fatally wounding a nineteen-year-old, Reginald L. Jenkins. Davis was charged with first-degree intentional homicide and Hills with aiding a felon.

McLean, Virginia—Narkey Keval Terry, 26, and Billy M. Canipe, Jr., also 26, were in morning rush hour on the George Washington Memorial Parkway. The specific events that led to the chase between the two at speeds approaching 80 mph are unclear, but Canipe was killed when his car crossed the median and slammed head-on into another vehicle, also killing its driver. They also killed a second innocent driver when Terry's Jeep hit and sheared off the rear of

Canipe's car, which hit a minivan. Terry was convicted on two counts of invol-
untary manslaughter and one count of reckless driving and sentenced to 10
years in prison.[3]

San Diego, California—Shawn Timothy Nelson, 35, felt his life was crumbling
around him. The divorced, recently evicted, alcoholic plumber took matters
into his own hands. He entered a National Guard Armory and stole a 57-ton
M-60 tank. Nelson traveled over 6 miles of residential road, running over 20
cars and vans, knocking down telephone lines and power poles, leaving 5,000
people without power. He ultimately crushed a telephone booth and several
fire hydrants. Fortunately, the tank's 105 mm cannon, 7.62 mm machine gun
and 12.7mm anti-aircraft gun were not loaded. Nelson's rampage came to a halt
when he stuck the tank on a concrete highway divider. At that point, four police officers leaped onto the tank, cut the hatch open with bolt cutters and
killed Nelson.[4]

Montgomery County, Maryland—Mike Tyson was convicted of second-
degree assault on Abimellec Saucedo, 62, and Dale Hardick, 50, after a traffic
accident in Gaithersburg, Maryland on August 31, 1998. He was sentenced to
one year in jail with another year suspended, two years probation, and 200
hours of community service after punching Saucedo in the face and kicking
Hardick in the groin. The incident happened after a chain-reaction collision
with Tyson's wife's Mercedes-Benz. Saucedo accidentally rear-ended Hardick's
vehicle, and in turn Hardick's vehicle nudged the Mercedes. Judge Stephen
Johnson called the action "potentially lethal road rage."[5]

These are only a few examples of a problem of great concern to law enforcement agencies, traffic engineers, emergency medical personnel, insurance companies, and almost the entire population of the United States.
"Road rage" is a popular term coined in the late 1980s to describe a type of
automobile driving behavior characterized by any display of aggression.
Most commonly, it is used to refer to more extreme acts of aggression, including personal attacks on other drivers, obscene gestures, verbal abuse,
throwing objects, and sometimes the physical assault of another vehicle or
driver. Journalists presumably derived "road rage" from the term "roid
rage," referring to the violent behavior sometimes associated with abuse of
steroids. The *Oxford Modern English Dictionary* recognized the term
"road rage" in 1996.

CURRENT HYSTERIA

The current focus on road rage is the product of several different factors,
not the least of which is media attention given to a series of studies

conducted in the United States and Great Britain. In 1995, the Road Safety Institute of the Automobile Association conducted a study on driving behavior in Great Britain and found that 88 percent of the motorists surveyed had experienced road rage incidents during the previous twelve months. The survey showed:

- 60 percent of drivers admitted losing their tempers while driving
- 62 percent owned up to tailgating aggressively
- 59 percent admitted headlight flashing
- 48 percent said they made obscene gestures
- 21 percent revealed deliberately obstructing other vehicles
- 16 percent admitted verbal abuse
- 1 percent admitted physical abuse.[6]

A 1997 study in Great Britain, the *LEX Report on Motoring*,[7] found that more than 2 million drivers had been victims of road rage. The severity of the incidents ranged from 1.4 million drivers reporting that they had to pull off the road to avoid an aggressive driver to approximately 130,000 who said other drivers had attacked them.

Recently, Gallup International questioned ten thousand motorists in sixteen European countries and found that in Britain 80.4 percent of drivers claim to have been victims of road rage, in Holland 78.1 percent, in Greece 76.6 percent, in France 70.0 percent, and in Germany 69.8 percent.[8]

Not to be outdone, the American Automobile Association's (AAA) Potomac Club commissioned a study in 1996 by the Gallup Organization to understand aggressive driving better. The study of drivers in the Washington, D.C., area revealed that aggressive drivers posed a greater perceived threat to highway safety (40 percent) than did drunk drivers (33 percent). The National Highway Traffic Safety Administration (NHTSA) reported similar results in 1997, with drivers designating excessive speed, aggressive driving, and unsafe lane changing as the major causes of accidents.

Aggressive driving was the number one concern (53 percent) among drivers in 1997 in the NHTSA survey. Interestingly, only 2 percent had mentioned aggressive driving in a similar 1994 survey. Taken together, these surveys, along with countless news reports and talk show interviews, suggest that something is happening on our roads, and that this something is quite troublesome to many drivers.

Western society revolves around automobiles. Americans, in particular, live in a culture defined by the cars they drive. In 1996, there were more than 206.3 million automobiles registered in the United States for a population of 265.2 million, being driven on more than 3.9 million miles of roads of varying composition, size, state of repair, and degree of governmental regulation. In Great Britain, there is one private car for every 3.5 people, while in the United States there is one car for every 1.3 persons. Over the past decade, there was a 17 percent increase in the number of cars in the United States, plus a 10 percent increase in the number of drivers. While these data are impressive, the real clincher is that besides the increase in the numbers of cars and drivers, Americans are also driving more. Since 1987, the average number of miles driven has increased 35 percent, while federal and state governments have constructed only 1 percent more roads.[9] This extraordinary increase in the sheer volume of cars, along with their increased use, has set the stage for an interesting case study of the intersection of our culture and biology.

WHAT IS "ROAD RAGE"?

The way that we define road rage and aggressive driving has important implications for our understanding. There is a lack of a consensus among those conducting the surveys about what is the "phenomenon of interest." According to Leon James, a professor of psychology at the University of Hawaii who has been doing research on road rage for the last fifteen years, habitual road rage is a persistent state of hostility when driving, demonstrated by acts on a continuum of escalating violence justified by righteous indignation.[10]

John A. Larson, a psychiatrist and director of the Institute of Stress Medicine, based in Connecticut, views road rage as the culmination of an escalating sequence of punitive behaviors meted out from one driver to another. What he calls "vigilante behavior" has four degrees of escalating expression: first, a gesture, curse, or grimace can be delivered as punishment; second, reciprocal exchanges of the same with decreased awareness of surroundings and impaired judgment; third, direct harassment of another driver through tailgating, high-beam flashing, retarding or blocking the progress of another driver (highway madness); and finally, intentionally injuring another driver or damaging a vehicle (road rage).[11]

Arnold Nerenberg, a California clinical psychologist, considers road rage to be a type of mental illness or anger disorder that the stress of driving can trigger. Road rage is a "driver reacting with anger at another driver; the anger is expressed overtly and communicated to the other." Nehrenberg describes three major traffic situations that can trigger road rage: first, feeling endangered, such as being cut off; second, being detained by other drivers who are driving too slowly; third, watching others breaking the rules of the road and thus feeling the need to retaliate.[12]

The legal system has also gotten into the act of defining road rage by identifying aggressive driving offenses and establishing penalties for them.[13] Arizona has passed a law creating the offense of aggressive driving, defining it as occurring when a driver speeds besides committing two other traffic violations, thereby creating an immediate hazard to others.

A bill in Washington describes aggressive driving as an offense in which a driver commits any two or more specific traffic violations within five consecutive miles in a way that "intimidates or threatens another person." Elsewhere, pending legislation defines aggressive driving as driving that intentionally creates a risk of harm or endangers the safety of others; operating a vehicle with a wanton or reckless disregard for another; driving with dangerous conduct contributing to the likelihood of a collision or evasive action by another; or driving in a deliberately discourteous and impatient manner.

One of the more intriguing ideas about road rage is found in legislation in Illinois, which envisions a continuum of traffic offenses that escalate from "moving violation" to "aggressive driving" to "reckless driving" to "road rage," the ultimate expression of aggressive driving. Some definitions mention the display of a weapon in a way that induces fear in another.

The central difficulty with all these is the identification of clear and unambiguous behavior that one can characterize as aggressive driving. It is likely that state legislatures, and ultimately law enforcement agencies and the judicial system, will have considerable difficulty dealing with these ambiguities in definition of the offense.

HISTORY OF ROAD RAGE

One of the frequently overlooked aspects of driving is the complexity of the task itself. Driving requires good vision, considerable hand-eye coordination, and careful coordination of vast muscle groups that simultaneously control the movement of our legs and feet, our arms and shoulders, and our heads. One must carry all these movements out with split-second precision, all the while moving much faster than even the most gifted sprinter. Driving an automobile is, contrary to the widely held popular belief of many teenagers, not a natural human activity, and it requires considerable training to master and coordinate all the necessary components. In fact, driving is a most unnatural human activity.

We fail to appreciate how unnatural the activity is until we put driving an automobile into an evolutionary perspective that includes our ancestral means of transportation. At best, humans can run as fast as 21.9 miles per hour,[14] but only for very short distances, while we can walk at a more leisurely pace (3.0 miles per hour) for considerable distances.[15]

It has been only in the last six thousand or so years that humans have had the luxury of relying on other modes of transportation than their own two legs. Archaeologists date the origin of horseback riding to approximately 4000 B.C.[16] Carriages, carts, and litters, drawn by or carried by domestic animals, have been a part of the human experience for centuries, but the "horseless carriage" is a recent invention. Gottlieb Daimler and Wilhelm Baybach built the first horseless carriage in Germany in 1887, followed by America's first internal-combustion motorcar, introduced in 1893 by brothers Charles and Frank Duryea.

From the beginning, there must have been tension between drivers of horseless carriages, drivers of horse-drawn vehicles, and pedestrians. In a book published in 1942, author Harry De Silva, a psychologist at Yale University, anticipated much of this chapter. He stated, "For generations human beings have been accustomed to relatively slow rates of travel. Now, although still endowed with nerves and muscles adjusted to slower tempos, they find themselves installed at the wheels of powerful high speed vehicles whose potentialities prove too much for them."[17]

The first reported automobile accident occurred on 30 May 1896, when Henry Wells of 70 Shamuck Avenue in Springfield, Massachusetts, collided with a bicycle ridden by Ebeling Thomas (459 W. 19th Street) on Western Boulevard in downtown Manhattan in the city of New York. Thomas suffered a fractured leg and was treated in the Manhattan Hospital. Wells spent the night in the West 125th Street police station.[18] We do not know if Mr. Wells was cited for his driving.

The British, however, have the dubious distinction of reporting the first automobile fatality. In August 1896, Mrs. Bridget Driscoll was killed in an accident near the Crystal Palace in South London. In 1899, Henry Bliss became the first American automobile fatality, and Mrs. Elma Wishmeir, a fifty-two-year-old worker in a Cleveland, Ohio, food-processing plant was the one millionth American to die on the highway when Herman Dengal from Dearborn, Michigan, ran over and killed her in 1977.[19] In the more than forty years since records have been collected, there have been approximately 1.9 million people killed on United States highways.[20]

We have a special way of dealing with automotive fatalities that is unlike how we deal with other kinds of mortal injury inflicted by one person on another. We call automotive deaths "accidents." This removes the sense of blame from the cars and the drivers. Lawyers frequently stress differences between traditional crimes and driving offenses. Driving offenses are characterized by the driver's liability for errors, miscalculations, or just really bad luck, as opposed to being wanton, intentional acts. Mistakes are the central part in the defense of most traffic accident deaths. Usually, no driving offense measures up to the popular idea of a criminal offense. It is well known in judicial circles that juries are reluctant to convict an individual of manslaughter who kills while driving, no matter how negligent.[21] For example, in Georgia, authorities define vehicular homicide as the unlawful killing of another with a vehicle, but, unlike murder, it does not require the intent to kill.

First-degree vehicular homicide is a felony.[22] Even the most egregious vehicular homicidal act is punishable by a maximum of only fifteen years of imprisonment (range two to fifteen years), and it is mandatory that time be served (one year minimum) only if the accused is a habitual violator. Second-degree vehicular homicide encompasses all other vehicular homicides without the intent to kill, such as failure to yield to oncoming traffic, speeding (not reckless driving), or driving too slowly. Second-degree vehicular homicide is a misdemeanor in Georgia and is punishable by imprisonment of up to one year and/or a fine of up to $1,000. However, a judge may suspend punishment or impose probation. In comparison, voluntary manslaughter is punishable by one to twenty years, involuntary manslaughter one to ten years, and aggravated assault one to twenty years.[23] I use Georgia as an example of the way that we, and Americans in particular, regulate driving an automobile differently from other potentially life-threatening behaviors. This is not to say that Georgia is better or worse than other states, but it typifies the cavalier manner in which American society deals with vehicular death.

The useless loss of life is the price we pay for our automobile culture. In Great Britain, when authorities hold individuals responsible for crashes by reckless driving, they fine those individuals £200, although the law allows for penalties up to five years in prison. In the United States, punishments are more stringent but vary considerably from state to state.

Seventeen states have introduced legislation that specifically deals with road rage, and others have bills under development. In 1998, nine states introduced aggressive driving bills, but, to date, states have enacted only two: Arizona's Aggressive Driving Bill (HB2311), which adds a section to the existing code and specifies penalties for various levels of offenses, and Virginia's Driver's Education Bill (HB896), which requires school driver education programs to include lessons to discourage aggressive driving. In states where legislation has been introduced, authorities define aggressive driving separately from other offenses. Most of the bills focus on penalties to distinguish violent driving acts (road rage) from other aggressive driving, by charge (felony vs. misdemeanor) and class, so that road rage is considered the highest level of aggressive driving. Bills also focus on education of the offender, including mandatory classes or inclusion in driver education classes. In addition several other states have legislation pending.[24]

It is interesting to consider how and why this disparity in our ideas about responsibility and driving behavior originated. Until after World War II, car ownership was a luxury, due to the high prices of automobiles and the associated maintenance costs. Automobile ownership inferred something about social class. Certainly this held true in England, and to a lesser extent in the United States. After World War II, automobiles became more widely available to the middle class. The wider availability of automobiles demanded closer regulation, and additional laws were enacted to control driving practices. This put those charged with enforcing the laws, the police and the courts, in direct conflict with those who had enjoyed driving largely unfettered by regulations, or at least by their enforcement. So those who had enjoyed the luxury of loosely regulated driving found themselves in a very different position. One can imagine that there was considerable pressure to modify the laws as applied to the wealthy.

Another possible explanation for the different treatment of driving offenders from other types of offenders is that the automobile is a recent cultural phenomenon, and rules governing it are not going to have the same degree of perceived validity as laws that govern other offenses.[25] In addition, there are powerful interest groups concerned that "criminalizing" driving offenses would infringe upon individual rights. The underlying assumption is that the motorist who breaks the law is not an antisocial criminal but a law-abiding citizen who will respond to "reasonable" treatment.[26]

Many judges have repeatedly shown that they do not regard the driving offender as criminal in the same sense as those who commit theft, robbery, murder, or rape. Those that do find the driving offender a criminal emphasize the danger to others and the disregard for the rights of others, as with drunk drivers and dangerous drivers. Many drivers, on the other hand, see laws that govern driving behavior as a threat to normally law-abiding citizens rather than as an attempt to control the criminal or potentially criminal element in the population.[27]

This preferential treatment of drivers of automobiles sets the stage for patterns of behavior that are irresponsible. If we behave aggressively and with criminal irresponsibility while operating a car, it is reasonable to believe we are unlikely to get caught. Even if authorities catch us, we are unlikely to be imprisoned. If imprisoned for, say, vehicular homicide, there is a greater likelihood of parole and certainly a sentence of shorter duration than if the circumstances of the homicide were different. Given the mechanical state and degree of sophistication of cars on the road today, it is safe to say that most crashes are not the result of mechanical failure but rather of human action.

The Long Island Protective Society (LIPS), incorporated in September 1902, was one of the most outspoken of the early antispeeding organizations. The society's legal counsel said,

> Our purpose is to enforce the speed law against the reckless drivers of automobiles and also those of fast horses. It is not a society antagonistic to automobiles. We recognize that the automobile is the twentieth century vehicle and it is with us to stay. Many of our members own and operate automobiles, but we are for a free highway and a safe highway and intend to harmonize the interests of the automobilists, the horse drivers, and the pedestrians.[28]

At the time of the incorporation of LIPS, only four states had passed special legislation dealing with automobiles. In town, regulations held automobiles at the top speed of horse-drawn vehicles, six to eight miles per hour in the business sections and ten to twelve miles per hour in other parts of cities. By 1906, most states had enacted some legislation that specified maximum speed limits of twenty to twenty-five miles per hour on the open highway. Given the number of horse-drawn vehicles and the state of the roads, this was about the maximum safe speed. Sentiment against speeding and reckless driving was especially strong in rural areas, where the automobile posed a danger to livestock and horse-drawn traffic and automobiles raised clouds of dust that ruined crops and invaded farmhouses and barns.[29]

Yet at the same time, around the turn of the century, politicians viewed automobiles as the solution to the country's transportation problems. They held the promise of relief for taxpayers from the expense of removing tons of animal waste daily from city streets plus the expense associated with miles of railroad track, overhead wires, endless tunnels, and the graft and corruption associated with urban mass-transit systems. In New York City at the turn of the century, horses deposited about 2.5 million pounds of manure and sixty thousand gallons of urine on the streets every day. Carcasses of overworked dray horses often clogged traffic because they dropped in their tracks during the summer heat or fell, stumbling on slippery pavement and breaking their legs. On average, street cleaners removed fifteen thousand dead horses from the streets of New York City each year. A 1908 estimate suggested that banning horses from New York City would save the city $100 million each year.[30]

One way to improve the quality of urban life, some suggested in the late nineteenth century, was to eliminate horses from the cities. The automobile heralded a new era, one that was free of the stresses and strains associated with urban living. The major source of stress in the urban environment, according to these motorcar enthusiasts, was the continuous noise of the iron wheels of horse-drawn vehicles on cobblestone streets. Although some carriages and bicycles had rubber tires, rubber tires did not become widespread until the development of the automobile. "The improvement in city conditions by the general adoption of the motorcar can hardly be overestimated. Streets clean, dustless and odorless, with light rubber tired vehicles moving swiftly and noiselessly over their smooth expanse, would eliminate a greater part of the nervousness, distraction, and strain of modern metropolitan life."[31]

This rosy picture of the automobile was not long-lived. By the 1920s, there was a growing climate of opinion that automobiles had not lived up to their initial promise and, rather than being a positive force in society, had become a threat to the tightly knit family and prevailing moral standards. Automobiles were beginning to be seen as the cause of, rather than the cure for, urban congestion. Merchants in the hearts of cities began to lament the introduction of the automobile because they perceived a loss of business due to traffic congestion. The mid-1920s saw increased concern over the safety of automobiles. The National Automobile Chamber of Commerce (NACC) Traffic and Safety Committee analyzed 401 automobile accident fatalities in 1924 and found that only 7 were due to mechanical defects, while environmental conditions (fog, pavement conditions) accounted for 92 and pedestrians were at fault in 150. The committee attributed the remaining 152 to the motorist, with excessive speed cited as the major cause in 48 and violations of "rules of the road" in 40.[32]

CURRENT STUDIES OF ROAD RAGE

There have been problems with aggressive driving since the introduction of wheeled vehicles, but few studies have attempted to deal with the problem scientifically. Meyer Parry wrote a book in 1968, *Aggression on the Road*,[33] in which he recognized the problem posed by aggressive driving. He characterized aggression as a personality characteristic of many individuals who were involved in accidents and observed that aggression could be latent or overt (e.g., a driver swearing at another when thwarted, as opposed to driving a vehicle straight at another person or vehicle) for perhaps the same reason. He did not use the term "road rage" but noted that, for some people, it was as if a magical transformation overcame them when they got behind the wheel. Individuals who were law-abiding and mild-mannered citizens became aggressive tyrants when in the driver's seat.

Another early study of driving behavior, while not recognizing road rage specifically, observed that many factors contribute to unsafe driving behavior. Besides fatigue, ill health, inattention, and intoxication, authorities also saw aggressive driving more than fifty years ago as a significant contributor to accidents. Another driver cutting in too sharply (after passing) or "hogging" the middle of the road, or backseat driving by occupants of one's own car, often shatter a driver's composure.[34]

Some drivers, by engaging in dangerous driving, intend to teach the offending drivers or passengers a lesson. Lack of self-esteem and feelings of personal inferiority may also lead to unsafe driving. To overcome the feelings of inferiority, drivers may aggressively pass others or refuse to allow themselves to be passed, by weaving in and out of a line of traffic to get to the front of the line or by speeding and engaging in other reckless acts. "The frustration of 'henpecked husbands' may similarly find an outlet in pugnacious behavior on the highway."[35]

Researchers conducted the most sophisticated early study of driving behavior in the late 1960s. It was the first true data-rich comparison of driving behavior on different continents. The authors concluded that road violence is one aspect of social violence. The greater the overall violence in society, they observed, the greater the aggression seen in driving behavior. Now this is not a particularly astonishing conclusion, but it is interesting because the researchers developed it in some detail using a sophisticated model of human behavior.[36]

In 1975, a study of driving behavior in the United States concluded that 30.1 percent of the males and 14.3 percent of the females in the study were guilty of aggressive driving, and when researchers considered age, 48 percent of the males younger than thirty drove aggressively. Such aggressive driving was not limited to the United States. Twenty-five percent of young male Scots (17–35) admitted to giving chase to drivers that had offended them, and 8 percent admitted to fighting with other motorists.[37] In 1971, Francis Whitlock noted that 85 percent of all traffic accidents in Great Britain were the result of aggressive driving.[38] In a study of drivers' reactions to pedestrians in a crosswalk, researchers in Holland found that 25.4 percent of the Dutch drivers tested were aggressive toward pedestrians, while similar studies in Denmark revealed that only 11.9 percent of the Danish drivers were aggressive when encountering a pedestrian.[39]

One of the most cited studies of road rage is a 1997 report commissioned by the AAA Foundation for Traffic Safety in Washington, D.C. The Louis Mizell Corporation, a company that maintains databases of crime reports in the United States and provides these data to various municipal governments, law enforcement agencies, and nonprofit organizations, conducted this study. It reported only the most extreme examples of aggressive driving, defined as "an incident in which an angry or impatient motorist or passenger intentionally injures or kills another motorist, passenger or pedestrian, or attempts to injure or kill another motorist, passenger or pedestrian, in response to a traffic altercation or grievance." The study also considered as aggressive driving occasions when a motorist intentionally drove a vehicle into a building or other structure. This survey is at the heart of most reports of aggressive driving in the United States.[40] There are several important points to keep in mind when considering the data provided in the Mizell report. First, the definition employed is the most extreme one that could be used. It does not include people killed or injured in drunk-driving or hit-and-run collisions. Moreover, the Mizell report is not a prospective study of the aggressive driving problem. It is a post hoc analysis of data collected for other purposes. At best it is a snapshot of the most violent of the aggressive driving incidents from the period of 1 January 1990 to 31 August 1996.[41]

While the Mizell study was not the only reason, it was a factor that led to the convening of a congressional hearing on 17 July 1997. Because of the study's impact, summarizing some of its findings is important. The data were collected over nearly seven years and include 10,037 incidents of violent aggressive driving (see definitions above). At least 218 men, women, and children had been killed because of these incidents and another 12,610 injured. The number of incidents had increased every year of the study. The Mizell report listed the following reasons (each is associated with at least twenty-five incidents that resulted in death or injury) given for the violent disputes: (a) arguing over a parking space;[42] (b) not allowing someone to pass; (c) playing the radio too loud; (d) giving someone the finger; (e) honking the horn; (f) following too closely; (g) failing to turn off an antitheft alarm; and (h) failing to signal. In approximately 37 percent of the 10,037 aggressive driving incidents, the perpetrators used a firearm, and in 35 percent a car was used as a weapon. Other popular weapons included fists, tire irons and jack handles, baseball bats, knives, ice picks, razors, swords, hurled projectiles (rocks, garbage, soda cans, partially eaten food), crowbars, lead pipes, two-by-four timbers, canes, tree limbs, hatchets, golf clubs, Mace, pepper spray, eggs, water pistols, and in one case a paint roller.[43]

The Mizell report also highlighted other causes of aggressive driving including domestic violence. In 322 instances in the Mizell sample, domestic violence (frequently jealousy) was the motivating factor behind the aggressive driving incident. The report attributed a similar number of aggressive driving incidents to racism. During the study there were no fewer than one hundred cases of men and women using their automobiles as battering rams, crashing through buildings and other property.[44] Aggressive drivers also plowed their vehicles into crowds of people and intentionally used vehicles to attack police or police vehicles. Carjackers have commandeered bulldozers, tractor-trailer trucks, and a tank for use in aggressive driving episodes.

The Mizell report characterizes drivers who engage in aggressive driving as typically young, poorly educated males who have criminal records, histories of violence, and drug/alcohol abuse.[45] Many of these individuals have suffered some emotional or professional crisis as a precipitator of the aggressive driving event. There is, however, considerable variation among the aggressive drivers, and noting differences within this composite portrait is important. Of the total 10,037 aggressive driving incidents, for example, males caused 9,096 (90.6 percent), while females caused only 4.1 percent.[46]

The Mizell report focused public attention on the problem, but a few critics have raised questions about the reality of the phenomenon in the first place. Michael Fumento, in an article in the *Atlantic Monthly*, noted that the Mizell report claimed that only 218 deaths were directly attributable to aggressive driving during a period when 290,000 people died in traffic accidents. Likewise, Mizell attributed 12,610 injuries to aggressive driving during a period when vehicles injured 23 million people. Furthermore, the data were not collected comprehensively but rather from approximately thirty newspapers, reports from sixteen police departments, and insurance company claim reports. While the increase in the number of cases was likely real, there was no attempt to test the claim statistically that they were on the increase. Without appropriate statistical tests, any causal relationship is suspect. The increases could have been due to sampling error and/or increased publicity about aggressive driving.

Fumento argues that there has always been a degree of aggression in driving. By putting a label on it, researchers have redefined it as a phenomenon and caused people to look for events to fit into the category.[47] Furthermore, critics of the Mizell report argue that, during the study period, deaths on American highways actually declined. The meaningful statistic, they say, is deaths per 100 million vehicle miles, which declined from 2.4 to 1.7 during the study. Similarly, the number of accidents declined from 151 to 141 per 100 million miles over the same period.

Recently, the AAA Foundation for Traffic Safety funded a study on road rage aimed at identifying interventions that might effectively modify the behavior of targeted drivers.[48] Investigators aimed the survey at the fifty largest metropolitan areas nationwide. They designed it to collect data on any aggressive driving programs that local agencies had set up, plus basic statistical data on the incidents themselves. The survey concluded that 54 percent of the respondents thought that road rage is a problem that needs addressing. Individuals that responded to this perceived need were asked about the most effective interventions against road rage. The most common response was increased law enforcement, but many recognized the need for behavior modification of the offending drivers. One of the most interesting parts of the survey was the detailed data on specific characteristics (time, day, weather conditions, current traffic conditions) of the events themselves. The study concluded that road rage was most likely to occur on a Friday afternoon, during peak travel times, and in fair weather. Inclement weather affected the incidence of road rage. Incidents occurred most frequently in the summer months and did not seem related to holiday travel. The most frequently encountered conditions were moderate traffic congestion combined with the use of alcohol or drugs.

A recent telephone survey conducted by the National Highway Traffic Safety Administration (NHTSA) of more than six thousand drivers of all ages found that 62 percent of the respondents said the behavior of another driver had been a threat to them during the last year. Also, age and gender were important factors in aggressive driving. Men were more likely to engage in unsafe driving, and young drivers were more prone to drive aggressively than older drivers.

Researchers have conducted other surveys in several cities in the United States, Canada, and the United Kingdom.[49] While they vary in sample size and methodology, the essential findings echo earlier studies. The overall pattern is clear for age and gender. That is, young males are the most likely to engage in unsafe driving behavior.

EXPLANATIONS FOR ROAD RAGE

Is aggressive driving an epidemic problem for society? The answer is complicated, and the most productive way to consider the problem is to rephrase the question. What we really are interested in is how to reduce the extraordinary number of fatalities on American highways. Aggressive driving has likely been a problem for the last six thousand years, and it became a problem when two vehicles had to share the same path or roadway. One can imagine the risk of being run down by a zealous Roman centurion or a team of draft horses pulling a beer wagon. The question that confronts us is not whether the aggressive driving we see today is something new, but whether it is simply the inevitable outcome of a conflict between our biology and our culture. While not a new problem, it is still a vexing one for scientists. Managing the expression of aggression is a challenge faced by all people in every culture, and the ways in which those challenges have been met are as variable as the people themselves.

The answer to the question of how to manage the expression of aggression and, ultimately, aggressive driving or road rage requires a bit of a digression into the ways in which science explains any phenomenon of interest. Philosophers of science have recognized that they can explain any observable phenomena at more than one level. For example, when confronted with a threatening stimulus, we might respond by fleeing or fighting. The sympathetic branch of the autonomic nervous system controls the fight-or-flight reaction. Responding to the threatening stimulus in the environment, an area of the brain called the hypothalamus activates the sympathetic nervous system, and that results in increased blood flow to the heart and the other viscera, the peripheral blood vessels, sweat glands, and also the eye and piloerector (hair-raising) muscles.[50] The increased cardiac output, altered body temperature and blood glucose, and dilation of the pupils permit rapid responses to the threatening stimulus. On one level, then, our response to the threat in our environment comes from the immediate physiological changes that occur within our body. We call explanations of this type *proximate* explanations.

On the other hand, we can explain our response to threatening stimuli as the outcome of our evolutionary history. In our evolutionary past, those individuals who responded vigorously enough to avoid being eaten by the predator, or who won in combat with other individuals, were the ones who survived and reproduced. In fact, those individuals with the most highly developed fight-or-flight response might have been the individuals most likely to survive and also outreproduce those possessing a less efficient physiological system. Having such an efficient system conferred an advantage on our ancestors, and under certain circumstances it still does today. This type of explanation relies on an understanding of the evolutionary history of the organism. Now let us see how we can apply these multiple levels of explanation to the subject of this chapter, road rage.

Proximate Explanations
Stress

Driving is a dangerous business. If you have ever been involved in an accident, you know that driving is dangerous. Nevertheless, even if you never have an accident, you must still consider that driving is a dangerous business, sometimes for reasons only indirectly related to the function of driving. For example, people consistently report the stressful effects of driving, particularly in congested, urban areas. No matter what the cause, stress has some well-defined and well-studied physiological concomitants. Driving has been found a significant stress inducer in several studies. In more than forty separate studies in several different countries, researchers have found increased cardiovascular disease morbidity and mortality among professional drivers.[51]

Professional driving requires a high level of attention that individuals must maintain for considerable periods to avoid serious and often fatal accidents. Driving is not so different from occupations that involve responsibility for the welfare of others and require constant vigilance. These occupations (e.g., ocean liner captain and air traffic controller) have been found significantly associated with increased risk of heart disease and hypertension. Not only does constant vigilance exact a toll on the human body; it exacerbates the costs under conditions of stress. Researchers have observed a systematically greater rise in blood pressure and cortisol (a stress hormone) in drivers faced with high time-pressure demands compared with those with low time-pressure demands.[52] Under conditions of stress, cortisol induces increases in blood glucose levels (remember our previous discussion of glucose in the context of Pima Indians), fatty acids, and also amino acids in the blood. Cortisol is necessary to form glucose from fats and proteins, and to provide ample supplies of glucose for the brain. Like much of human physiology, a little is good, but a lot is not necessarily better. Too much cortisol production can inhibit bone formation, depress the immune system, and alter cardiovascular, neural, and gastrointestinal function.

Given these adverse physiological effects of driving on those who do it for a living, it is not too great a stretch to imagine that those of us who do not drive professionally are also subject to many of these same stresses, admittedly at a reduced intensity. These physiological concomitants of driving behavior and, in particular, the increased stress of driving ought to alert us that driving is an activity in which our biology and our culture are potentially in conflict, and this can lead to aggressive driving.

Early Experience

Another possible explanation for aggressive driving relies on the learning capabilities of modern humans. Investigators have made much about the effects of early exposure to images of graphic aggressive behavior on the social development of children. In a broader context, there is a rich literature on the importance of early experience on the fundamental underlying substrates of behavior. Harry Harlow, in a classic series of experiments on rhesus monkeys, showed the importance of comforting physical contact to normal social development (e.g., mothers holding infants).[53] He related deficits in the early rearing environment to patterns of antisocial behavior in adult monkeys. This animal research is relevant because it illustrates a relationship of early experience to development that may also apply to humans.

Indeed, some researchers contend that early exposure to aggressive and/or antisocial behavior can profoundly affect adult patterns of human behavior.[54] In particular, many psychologists interested in road rage contend that aggressive driving is the adult expression of behaviors acquired in childhood. Children today are reared in a car culture that condones irate expressions of emotion and behavior as a normal feature of driving a car. Parents teach children the lesson that once they are in the car it is acceptable to show aggression, lose control, and use language that under other circumstances they do not allow. Two case studies illustrate the problem:

> Suburban Washington, D.C.—Anne, a forty-year-old mother of three, backed her two and a half ton Chevy Suburban out of her driveway. The clock on the dash read 2:16. She had fourteen minutes to make it to her daughter's game. Within a block of her house she was approaching forty miles per hour, taking stop signs as suggestions rather than the law. She had a lot on her mind and not even thinking about other cars, Anne cheerfully admits as she lays on the horn. An oldster in an econo-box ahead of her has made a near-fatal mistake of slowing at an intersection with no stop sign or traffic signal. Anne swears and peels off around him. "Make him move over," shout the kids from the back seat, as she bears down on a sluggard driving 55 mph in the fast lane. She flashes her headlights and the unlucky target moves into the center lane. Back in town, Anne specializes in near misses. "Jeez, I almost hit that woman," she chirps, swinging the Suburban into the right lane to pass a car turning left at an intersection. She makes the game at 2:32. "I don't think that I am an aggressive driver, but there are a lot of bad ones out there," she says.[55]

> Westport, Connecticut—Carol Rodriguez was driving her daughter to school. They were late, as usual, and Rodriguez gripped the steering wheel tightly and stepped on the gas. They came up behind an older woman driving the speed

limit. Suddenly, her 14 year-old angrily yelled out, "Come on, get moving! Whaddya doing?"[56]

While researchers have not demonstrated that children who are exposed to aggressive driving will themselves turn out to be aggressive drivers, many studies point to the influence of early socialization agents on later human behavior. Given the exposure to violence that is a part of American society, exercising restraint in expressing aggression and antisocial behavior in the presence of children seems prudent, particularly while in an automobile.

Anonymity

Modern automobiles are unparalleled in their comfort, convenience, and overall ease of use. Air conditioning, power heated and cooled leather seats, power steering, power brakes, sound-deadening foam and padding, audiophile-quality stereo systems, televisions, electronic games, Internet access, infrared night vision, and Global Positioning Satellite (GPS) systems make our cars more comfortable than many homes. Couple these conveniences with the imposing size of many vehicles on the American road and you have the potential for some rather bizarre behavior.

In most urban areas, you know only a small fraction of the people you meet on your daily travels to work, to school, to the grocery store, and so on. That sense of anonymity may be a factor in our aggressive driving. There is the feeling that you are unlikely ever to see these people again, so why is it necessary to accord them even the most rudimentary of courtesies? In the comfort and safety of a car, it is easy to imagine how one might be tempted to "punish" errant drivers if they are no more than unknown faces behind the wheel. Now I am not suggesting that anonymity alone will cause aggressive driving. Yet there is good evidence that under certain conditions anonymity is one variable that can affect the expression of aggression.[57]

Road Congestion

Everyone knows the feeling of being stuck in traffic on a major expressway. You think to yourself that surely there must be some major accident ahead, and death and destruction must abound given your glacier-like pace. You patiently move slowly along, scanning the horizon for signs of the accident, but without success. Traffic is simply so congested that the six-lane freeway resembles a skinny parking lot many miles long.[58]

Studies have shown, however, that the slow crawl of congested traffic is not where most aggressive driving occurs. The time for really aggressive driving comes not during impenetrable gridlock but just before complete saturation, when traffic density is high but cars are still moving briskly. A recent report issued by the AAA Foundation for Traffic Safety concluded that while the density of traffic was a contributor to aggressive driving, road rage was actually most likely to occur under conditions of moderate to light congestion. Only a little more than 20 percent of the incidents reported occurred in heavily congested traffic.[59]

Well, there is good news and bad news. The good news is the completion of a study from the Texas Transportation Institute at Texas A&M University that reviews roadway congestion in the sixty-eight largest urban areas of the United States since 1987 and quantifies the problem in a scientific manner.[60] Researchers calculated a roadway congestion index (RCI), a measure of vehicle density on major roadways. An RCI of 1.0 means that a route takes no longer to drive during peak periods of use than when one has less crowded conditions. An RCI greater than 1 says that, during peak periods of use, routes take longer than when one has less crowded conditions. The greater the RCI, the longer routes take under congested conditions.[61] This study is not news, but the quantification of the delay is new and important.

Now the bad news: during the study period, in one-fifth of the urban areas sampled, drivers experienced trips that took 30 percent longer due to congestion. In more than one-third of the areas studied, average delays per driver exceeded one workweek per year. Los Angeles, Seattle, San Francisco, Washington, and Chicago top the list of cities with the greatest delays. In Los Angeles, an average trip taken at peak congestion time takes 51 percent longer than during off-peak times.

The report shows something that we all have known, which is that congestion increases with city size. However, the rate of the growth in the economy may influence rapid increases in the RCI more than the size of the area. Significant increases in the number of residents and jobs usually occur before government expands the transportation system.

Obviously, the data presented in the Texas Transportation Report simply confirm what we already knew. For most of us living in urban areas, traffic congestion is bad and getting worse. There are approximately 3.9 million miles of roads in the United States today, only 1 percent more than in 1987. On the other hand, the number of vehicle miles driven has increased 35 percent during the same period. The number of cars has also increased 27 percent during the same period, but the rate of population increase since 1987 was about 10 percent.

Beyond the increases that are directly attributable to increases in population, there are extraordinary increases in both the number of cars on the road and the amount that we drive them. While not the only predictor of aggressive driving, traffic congestion is clearly not likely to decrease, at least in the short term, so anticipating continued high levels of frustration among commuters in most urban areas is reasonable.[62]

Cultural Values

Styles of driving and attitudes about driving in some ways reflect the cultural values of society. People have said that the British are more polite drivers than Americans, but a recent poll found that British motorists suffer from road rage more than other European drivers. Germans drivers are less polite than the British, but Germans make much more efficient use of the road. German drivers accept the risks of driving more than the British and actually take more risks than British drivers but experience less road rage.[63]

Specific driving rules also vary across cultures. Even within a culture, or for that matter within a single urban area, there can be considerable variation in driving behavior. One of the classic regional differences in the United States is what people deem appropriate behavior when confronted with a yellow traffic signal. In some areas, the unwritten rule is something like "three on yellow." Translated, that means three cars are allowed through an intersection after the traffic signal has changed to yellow. In other areas, a yellow light means the same thing as a red one. In those "stop on yellow" parts of the country, the local unwritten rule if you are stationary at an intersection is that you can start into the intersection when the opposing traffic signal turns yellow. Imagine the potential for trouble if we mix some "three on yellow" drivers with a few "stop on yellows." Add in a few more for whom English is a second language, and you have a condition frequently seen on many roads in the United States.

What is intriguing is the way in which our driving reflects the nature of society. John A. Larson, psychiatrist and authority on road rage, characterizes the aggressive driver as one who feels frustrated in the pursuit of widely accepted cultural goals. Getting to one's destination in the shortest time possible is one of the highest-priority goals of individuals who engage in road rage. They perceive time spent in transit as wasted. The aggressive driver who punishes another driver is (sees himself or herself as) the victim. Although the aggressive driver recognizes that such behavior may be inappropriate, he or she still feels justified in taking action to deal with the threat to the attainment of acceptable goals. It is likely that aggressive driving is more frequent in areas where there is a premium placed on time optimization, rather than in areas where time is something to be enjoyed and savored. One can imagine that huge cultural differences in the perception of time would have a significant impact on the amount of aggressive driving.

The Faster the Better. Americans have largely accepted the idea that time is an adversary and must be overcome. I am not going to venture into a discussion of why this is the case, but anthropologists, philosophers, and others have been interested in this attitude for a long time.[64] While driving, motorists can routinely be seen reading, applying makeup, talking on the telephone, dictating memos, eating lunch, or more distressingly, some or all of the above simultaneously. But above and beyond such small victories over wasted time, the goal of travel is to arrive at one's destination in the shortest possible time. Aggressive drivers fully embody this cultural value. Once on the road, aggressive drivers set rigid expectations for themselves and often seem as if their self-esteem is dependent upon reaching the destination in a time equal to or less than the established goal. They see any delays as impeding the achievement of the goal. There is nothing intrinsically wrong with setting time goals and making reasonable efforts to achieve them, but when the goal of reaching the destination in the specified time takes precedence over passenger safety, the safety of other drivers, and exercising caution in the face of adverse driving conditions, something is amiss.

Many aspects of American culture attach value to accomplishing goals as quickly as possible. Commercial radio and television preach speed of information delivery: headline news, news in the first eleven minutes, sound bites from around the around world in thirty seconds—to mention just a few. While some automobile commercials and advertisements promote safety, the vast majority promote speed. A primary basis for comparison for many car shoppers, particularly young males (the age group that is most likely to engage in risk taking), is the elapsed time from zero to sixty miles per hour, lateral acceleration, quarter-mile time, top speed, and so forth. Car design emphasizes speed, "racy looks." Movies rely on speed and car chases; one even has the title *Speed*.[65] To be the fastest, whether you are a race car driver, a marathon runner, or a halfback in football, is the goal. The whole notion of "fastest is best" has taken on the trappings of a moral imperative for American society. Same-day mail delivery, faxes, voice mail, e-mail, drive-in drugstores, drive-in liquor stores, "minute" marts, drive-in wedding chapels, and "fast" food all place a premium on speed.

Driving Is an Inalienable Right? Given that there is almost one vehicle in the United States for every man, woman, and child, it is not surprising that Americans take driving for granted. Many define their state of physical and mental well-being based on the ability to drive. It is a landmark event for a teenager to get a driver's license, but it is also a landmark event in one's life to give it up. In this respect, the automobile is an equalizer. To be sure, different types of automobiles signal differences in socioeconomic status, but the real distinction is in owning one or not. Given that the car culture has so completely immersed Americans, it is easy to see how, rather than being a privilege, most consider driving a right. While not specifically covered in the Constitution, the freedom inherent in the United States speaks directly to one's right to drive. One of the fundamental bases of economic growth and development in the United States in the twentieth century was the widespread availability of automobiles. Rather than having to rely on walking or public transportation, which placed real constraints on how Americans developed cities, entrepreneurs have been free to spread out farther and farther in the quest for convenient parking.

The emphasis on driving, however, also brought commuter congestion. Part of the problem was that jobs shifted from the cities into the suburbs. Initially designed as residential suburbs with narrow roads, "edge cities" have sprung up all across the United States. Suburb-to-suburb commuting accounts for 44 percent of all metropolitan traffic, while suburb-to-downtown commuting only 20 percent.[66] Joel Garreau, demographer and author of *Edge City*, says that workers breaking for lunch are creating what is rapidly becoming a third rush hour.[67] Because Americans have designed most mass transit systems like spokes radiating from a central hub, centering in the cities and branching out to the suburbs, they are not useful in getting from point A to point B in an edge city, nor are they useful in commuting between edge cities. It is no wonder that few people in the United States rely on mass transit to go to work. In 1969, 82.7 percent drove to work, while in 1990, 91.4 percent did.

Given that Americans have so intimately intertwined their economic well-being with the automobile, it is no wonder that the conditions for commuting are likely only to get worse. Unless there is a major change in the driving ethos in the United States or there are significant changes in patterns of driving, Americans in particular will find that the problem of aggressive drivers is here to stay.

Value of Competitive and Aggressive Behavior. Americans are competitive. The country was founded on economic competition, and being successful at competitive encounters is highly valued in American society. Vince Lombardi, legendary coach of the professional football Green Bay Packers, is known for valuing competitiveness and victory at any cost. He is reported to have said during his opening remarks at his first training camp in 1959, "Winning isn't everything, it is the only thing."[68]

Americans have the idea that second place is nothing more than the first loser. To win, one must have opponents. Aggressive drivers see themselves in competition with other drivers. The other drivers are the enemy that must be defeated. It is the competitive nature of many American drivers that allows them to engage in contests that, if acted out in an automobile, can have fatal consequences. Again, there is nothing wrong with competition, but when the outcome of the competitive encounter assumes larger-than-life proportions, there is something drastically out of balance.

The cultural preoccupation with winning has become the principle for much of the behavior of Americans. In many day-to-day interactions, the competitive dynamic supersedes a fundamental concern for others. This competition mentality often manifests itself as redefining what might be normal patterns of cooperative behavior into "win/lose" situations. For aggressive drivers, this win/lose system comes into play when one driver perceives aggressive behavior being directed toward him. This aggressive behavior can take a variety of forms including horn honking, tailgating, or headlight flashing. This initial aggressive behavior quickly transforms the interaction between two drivers into a competitive encounter. The encounter becomes intensely personal, and the self-esteem of both drivers is on the line and being contested. Once this happens and the interaction escalates into one of winning or losing, then the possibility of a lethal encounter becomes very real.

According to some, certain personality types are at high risk for engaging in aggressive driving. Young males often strongly identify with sports and sports heroes and are, by their evolutionary heritage, more likely to take risks than any other segment of society. They fully internalize the competitive "life is a game" mentality. Males more often exhibit a type A personality than do females. Medical researchers have identified these type A individuals as "coronary-prone." The type A person has a cluster of attitudes that include time urgency, competitiveness, and hostility toward anyone perceived to be delaying or blocking his progress, plus a pattern of reactivity that appears to interact with the stresses of driving in potentially damaging ways.

Others at risk for aggressive driving are those with displaced anger and projected rage. These individuals may have a circumscribed anger problem specifically related to the stresses of driving (e.g., traffic jams, construction delays, parking space piracies), or they may suffer anger from unresolved conflicts in other parts of their lives. Another personality type implicated in aggressive driving is the passive-aggressive (Jekyll and Hyde) type. These are individuals who are prone to displace their anger or rage. Often seen as meek and unassertive in nondriving situations, passive-aggressive individuals become hostile and aggressive when confronted with perceived driving conflicts.

Finally, there is what some have called the "polite rule enforcer." Contrary to the typical case with other aggressive drivers, these individuals are more likely to be female. An overreaction to perceived transgressions and rules characterizes the "enforcers" of driving etiquette (e.g., failure to yield the right of way or going out of the proper order at a four-way stop). The important point here is that there exists an identifiable suite of personality types that are at risk for aggressive driving.

To add insult to injury, aggressive drivers not only feel that they have the right to win the encounter but also, typically, that it is their duty to punish the transgressor. Thus, they see driving as competition with a clear winner and loser. Moreover, aggressive drivers feel justified in meting out some type of retribution to others who have impeded the attainment of their driving goals. It is this righteous indignation that characterizes much of what we call aggressive driving.

Three Ways of Doing Anything: Right Way, Wrong Way, and My Way. Americans often exhibit a conspicuous lack of tolerance for individuals who look or behave differently from themselves. Many immediately ascribe anyone who does not drive the way they want to the "asshole driver" category. This contemptuous attitude becomes expressed through discourteous and potentially dangerous driving behavior. The aggressive driver is always alert for any sign of "poor" driving. In a sense, the aggressive driver is looking for trouble and all too often finds it.

The size and power of American automobiles fuel this narcissistic behavior on the road largely because of the anonymity they afford. Americans have precious little information about the circumstances or motivations of other drivers. The "asshole" that cuts off an aggressive driver may be pregnant, trying to get to a hospital while simultaneously trying to comfort a two-year-old in the infant seat behind her. Darkened windows further increase anonymity, so that what one may interpret as an aggressive act committed by the other driver, given the lack of information, may not be that at all. That said, once the "victim" has interpreted the behavior as aggressive, then responding to it is okay.

Drivers sometimes use the type or make of car one is driving as information about intention and motivation. Individuals react to the "personality" of the vehicle, not the person who is in it. There has been much made in advertising about the name of a car and what it says about the owner. It is true that automobile manufacturers use names that conjure up thoughts of aggressiveness, competition, or speed.

Given the enormous amount of automobile advertising, it is not surprising that Americans quickly make inferences about a driver from the type of car driven. What kind of person drives a Mustang, Barracuda, Viper, Range Rover, Taurus, Cobra, or a Stingray? Is it any wonder that drivers of Accords, Escorts, Neons, or Foci may be characterized as nonaggressive? What would be the reaction to the now defunct Studebaker Dictator, Chrysler Imperial, Dodge Demon, Oldsmobile Starfire, or Buick Roadmaster? When confronted with the lack of substantive information about a driver's motivations, the aggressive driver makes decisions that can have life-threatening consequences.

Psychiatric Disorder

Arnold Nerenberg, a clinical psychologist in Whittier, California, considers road rage and aggressive driving a type of mental illness or anger disorder that the pressures of driving can trigger. Nerenberg sees a familial connection to the disorder. A child growing up with someone who routinely expresses road rage creates a predisposition to develop the same behaviors. Nerenberg, who has appointed himself "America's Road Rage Therapist," estimates that 60 million drivers in the United States experience road rage on the average of four times per year, producing 250 million accidents per year. It appears, however, that Nerenberg has not tabulated his data correctly, for he has also said that 53 percent of our population has a road rage disorder and 1.78 billion episodes occur each year. This is because, on average, drivers manifest road rage twenty-seven times a year and we have about 125 million cars on the road.[69] One is tempted to wonder if Nerenberg might be inflating the estimates for personal gain.

In any event, Nerenberg characterizes two different levels of road rage. At the nonfelonious level, it can include screaming, making obscene gestures, beeping the horn, flashing headlights, hostile stares, and spitting. At the felonious level, it escalates into ramming another car, firing guns, and chasing after other drivers with the intent to harm them. Nerenberg has proposed that road rage be included in the *Diagnostic and Statistical Manual of Mental Disorders*, the clinical authority on diagnosis of psychopathology published by the American Psychiatric Association.[70]

Ultimate Explanations

This chapter has focused on an aspect of human behavior that has gotten much attention from the media in recent years. At best, the attempts to explain aggressive driving or road rage have relied on proximate or immediate causes. A problem in any discussion of this kind is, of course, that many factors influence behavior on the level of the individual and we can invoke many of them to explain whatever aspect of human behavior is of interest. To understand any aspect of human behavior fully, we must seek explanations on multiple levels. Here, I focus on answers not to the question of how aggressive driving occurs but to the larger question of why it occurs.

Let me rephrase the question in terms of modern evolutionary biology. Is there some aspect of human aggressive behavior that in our evolutionary past could have been fitness enhancing? The answer is obvious. Clearly, in our evolutionary past, natural selection favored aggressive behavior under certain conditions. Individuals who acted aggressively in appropriate situations enjoyed greater fitness than did those who were more pacific. Let me be clear. This does not mean that all of our aggressive behavior is based in our evolutionary past and therefore our genes, nor does it mean that aggression is always a fitness-enhancing behavior. There are times when admitting defeat and living to fight another day is best.

Some deeply rooted aspects of human behavior that were fitness enhancing in the evolutionary past may actually be deleterious in the modern environment. Certain aspects of our modern environment may trigger behaviors that served our ancestors quite well, and still may do so for us under certain circumstances, but our modern environment may also lead to behaviors that now run counter to our best biological interests.

Territorial Imperative

Studies of animal behavior have the potential to illuminate various adaptive solutions and constraints on the way that animals solve the fundamental problems of life. Insights gained from the study of animals have helped us understand more fully the aspects of our behavior that we share with other animals due to phylogenetic heritage, as well as those uniquely human. People have argued that humans, like some other animal and bird species, are territorial.[71] Territoriality is defined as the aggressive maintenance of an economically defendable space against intruders.[72] The study of the human tendency to attempt to control the space that immediately surrounds the individual and how local ecological factors as well as cultural factors may influence the spacing system is called proxemics.

A review of human proxemics is well beyond the scope of this chapter, but it is important to remember that we each have an area of personal space that, if violated, makes us feel uncomfortable at best. Consider the case of riding on a crowded subway. Personal space is virtually zero as passengers are pressed cheek to jowl. Given the opportunity, people will spread apart

and reestablish their personal space. In modern society, rarely do we resort
to physical violence to establish personal space. Much more often we rely
on nonverbal cues or so-called body language.

Some have argued that the car is indeed an extension of our personal
space and attempts to invade our car space may be met with the same reaction as overcrowding on a subway.[73] While we might respond to pushing
or crowding on the subway by simply pushing back, or even turning away
from the perpetrator, that is very difficult in an automobile. The violation of
car space might be nothing more than someone approaching too close to
our rear bumper while stopped at a traffic light. As the saying goes, "if you
can't see their headlights, they are too close." The response to this invasion
of perceived personal space may be to step on the brakes or inch forward,
away from the intruder.

If in motion, an intrusion into our car space could take the form of
someone abruptly changing lanes without a signal. This kind of violation of
car space is likely to elicit a response aimed at reestablishing the appropriate space. Honking the horn and blinking the lights are common tactics to
alert another driver that they have intruded on your car space. The real difference in attempts to maintain car space versus personal space is that the
range of nonverbal cues that we might use in the subway are not available
to us while driving an automobile. A very limited number of nonverbal
communication signals are available to us, but giving another driver a nonverbal signal, the middle finger salute, for example, often has the effect of
escalating the confrontation. The violated individual may feel that simply
communicating displeasure is not enough, and it becomes the duty of the
injured party to "teach that bastard a lesson."

Aggressive territorial defense is found among animals and birds, but the
conflict is typically resolved through nonviolent means. Our nonverbal signals have evolved because they are an effective means of communicating
information with the least likelihood of physical aggression.[74] Simply put,
we do not have any way to signal submission or an apology for our inappropriate behavior while driving a car. As a matter of fact, any attempt to
signal contrition is likely to bring about further aggression.

So what if this evolutionary hypothesis of territoriality is behind our aggressive driving? What if people in one region were all to agree that they
would signal "sorry" by raising a hand in front of the face, with palm toward
the face, the forefinger and middle finger extended to form a V shape, and
the other two fingers and thumb fully bent? However, individual cultural
biases might hinder widespread acceptance. Instead of "sorry," what if the
V said something about your mother's preference in men? Rather than be-

ing reduced, aggressive driving might then actually increase, depending on the other driver's cultural heritage.[75]

If anyone doubts that many Americans feel something about their car space, try this. While riding a bicycle in traffic, stop at an intersection immediately adjacent to a stopped car. Rather than simply put a foot on the ground to balance yourself, put your hand on the roof of the adjacent car. But be prepared to make a quick right turn, as this behavior is likely to precipitate a strong negative response from the driver.

Explanations of human behavior that rely on the tight genetic control of our behavior, like explanations of the genetic inevitability of aggression or the difficulty that many men have remaining strictly monogamous, miss the point of much of human behavior. Human behavior is not as tightly controlled by our genes as is the behavior of some other species. The study of the behavioral genetics of a creature that we often see as an obnoxious little pest, the fruit fly (*Drosophila melanogaster* or my personal favorite *Drosophila pseudoobscura*), has shown us how powerful genes can be in controlling behavior. For the fruit fly, it is as if genes have behavior on a leash, and a very short one at that. So how does a discussion of the behavioral genetics of *Drosophila* relate to our initial question about aggressive driving? I am not sure that it does directly, except that it is naïve to think that human behavior can be completely explained by genes. Our behavior reflects not only our genes but also the culture that surrounds us. We rely heavily on learning as a way of surviving in our environment. Our culture teaches us all sorts of lessons about how to interact with others, what is appropriate and what is inappropriate behavior. We are not fruit flies. Our behavior is a complex dialogue between our genes and our culture.

Maybe I can illustrate the relationship of genes and culture by way of analogy. Many people play chess, a game that arose in India in the fifth century and spread rapidly throughout the Old World by the thirteenth century. The game has been variously described as the least exhaustible of pastimes, mental torture, and something for which life is too short.[76] It is a complicated game with strategies dictated in a dynamic fashion, being played out with each move of your opponent. For all its complexity, the rules are relatively simple. I dare say that, for most, a full understanding of the rules (although not all the strategies) can be gained in a matter of hours. Does this knowledge of the rules equip you to play an international grand master?[77] The answer is no, obviously.

To fully grasp the subtleties of the game, one must devote years to intensive play, and then only a few (only about four hundred today) reach the expertise level of an international grand master. Now let us take this analogy a bit further. Computer geeks have been engaging in elaborate chess games for years, trying to develop a computer that is able to play at the level of a grand master.[78] The question facing the computer wizards was how do you teach someone to play chess? Of course, a number of training strategies have been tried, but the one that seemed to work best was a strategy of simplicity. Experts quickly realized that they would not be able to program all the possible moves that could be made on a chessboard, so what they did was to program a basic set of rules and let the computer learn while it was playing.

To me, the relationship between our genes and our culture is like computer chess. Our genes set the rules of the game, but they do not determine every move we make. We constantly evaluate the local environment and modify our behavior accordingly. Without this flexibility in behavior, humans would not have become the most successful mammal on the planet. Can we account for human aggressiveness as a simple outcome of our genetic heritage? I think not. Human behavior is the complex interplay between the rules that are set out by our genes and learned patterns of behavior that we use to deal with the environment around us.

This means that it is highly unlikely that aggressive driving or road rage is the simple outcome of ancient inherited behavioral programs. We are not in the iron grip of our aggressive, territorial nature. We are playing out a constantly changing game with our environment that is loosely guided by our genetic heritage.

Discordance Hypothesis

From my perspective, a much more likely explanation, and one that is consistent with the central theme of this book, is called the discordance hypothesis. As you have seen, the discordance hypothesis has some merit in explaining the seemingly paradoxical behavior of modern humans to engage in behaviors that are clearly detrimental to health and fitness. In spite of massive amounts of scientific data on the negative effects of carbon monoxide on lung function and the addictive effects of nicotine, millions of people light up cigarettes every day. In spite of the risks, many people knowingly engage in potentially life-threatening behavior while driving a car. The lack of what we would call rational thinking seems astonishing. What could account for this seemingly paradoxical behavior?

The aggressive behavior that is exhibited on the roads in the United States may be an example of the discordance between our biology and our behavior. Nothing in our past specifically equips us for driving a three-thousand-pound machine while negotiating a twisting and turning road with several more of these three-thousand-pound motorized boxes within a few feet of us. There may be, however, patterns of behavior elicited by immediate environmental circumstances that are really a part of our evolutionary survival kit. I am not saying that the erratic and dangerous behavior that we called road rage is a direct expression of our distant evolutionary past. It may be that road rage is a complex suite of behaviors that are largely learned but that also trigger something from our distant past. Nor am I saying that road rage is a genetically preprogrammed behavior; rather it is a behavior that taps responses that were adaptive in our evolutionary past but are clearly detrimental today.

I am suggesting that in many ways we are very poorly equipped to deal with our modern environment. Many of the problems we have today are the result of a physiology that was adapted for life fifty thousand or so years ago. Yet we find ourselves in an environment that is changing at an ever increasing pace. We are remarkably poorly equipped to deal with the stresses of modern automobile travel. Our stress physiology, for example, has equipped us for dealing with life-threatening situations as we might have encountered them in the past. In no way are we equipped to deal with many of the stresses in today's environment. It is this discordance between our evolved physiology and our environment that leaves us in such a cruel bind. But is this simply an excuse for our behavior? Can those who have caused the death of another driver simply say that it was the lack of fit between their physiology and the environment that made them do it? I think not. An understanding of the discordance between our physiology and the environment should be seen as an opportunity for changing human behavior. Armed with this evolutionary perspective, we can see features of both our environment and our behavior that can be modified.

From my perspective as an evolutionary anthropologist, it is remarkable that we do not see more cases of lethal aggression on our highways. We place ourselves in situations that we know are likely to be stressful, we isolate ourselves from contact with those around us, virtually eliminating the possibility of conflict resolution "the old-fashioned way," we arm ourselves with three-thousand-pound projectiles, and, on top of all that, we carry firearms for protection.

In the Mizell study, to which I have referred numerous times in this chapter, it is reported that, in 35 percent of the incidents of aggressive driving, the vehicle itself was used as a weapon, while in 37 percent a firearm was used. It seems to me that, given the highly volatile nature of driving anyway and the inevitable possibility that vehicles themselves will be used as weapons, it makes no sense to further prepare drivers to resolve conflict by arming them with lethal weapons that can easily be used while sitting in the relative safety of a car. Handguns raise the stakes in aggressive driving encounters. According to the National Rifle Association's Institute for Legislative Action, in over three-fourths of the states in the United States residents can legally carry a loaded handgun in a car. In one-third of those states where carrying a loaded weapon is legal, laws do not require a permit if the weapon is not concealed. That means that in thirteen states you can legally carry a loaded handgun on the dash, the front seat, in your lap, and, in at least one state, in the console or glove box without any official permit. Given this state of affairs, I am surprised that there are not more murders on the nation's highways.

So what do evolutionary theory and the discordance hypothesis offer as a solution to aggressive driving and road rage? Unfortunately, there are no easy answers, nor are there quick fixes; however, it might be useful to think about how we might manage in a more positive way the stress imposed on us by our culture.

RECOMMENDATIONS

Can anything be done to decrease the terrible cost that aggressive driving inflicts upon society? An evolutionary perspective helps contextualize the aggression we see, not condone it. If we can understand at least a part of the motivation for engaging in any behavior, then it is possible to go about devising strategies to change it. There are a number of approaches that can be taken, but since the problem is so poorly studied, it is premature to suggest that any single measure will be more effective than another.

Better Enforcement

Ricardo Martinez, head of the National Highway Traffic Safety Administration, in his testimony before the United States House of Representatives Subcommittee on Surface Transportation, focused on several things that can be done to combat aggressive driving. One of these was better enforcement. He noted that aggressive drivers must be held accountable for their actions and the best way to do so would be vigorous enforcement of existing traffic laws. Knowledge that there will be swift and severe penalties for violation of laws is an effective way to reduce aggressive driving. Several states have enacted changes or have changes pending in the laws governing aggressive driving, as well as special highway patrol units assigned specifically to counter aggressive driving. Other measures he recommended include increased use of high-tech speed detection devices and communications technology.

Martinez also called for more vigorous prosecution of offenders. Aggressive driving should be taken as seriously as drunk driving if we are to control it. Charges should be substantial, including, where appropriate, reckless endangerment and vehicular homicide. Courts must send a clear and consistent message to the driving public that aggressive driving will not be tolerated, according to Martinez.

Now as a student of human behavior, I predict that some of Martinez's recommendations would probably be effective, at least in the short run. Yet it is clear that there will never be sufficient enforcement of existing laws to stop all aggressive driving. Moreover, all the changes in existing laws will do little to change driving behavior if they are not enforced and the courts do not mete out swift and decisive punishment. One of the greatest hindrances to the use of punishment as a means to change behavior is the time delay between the offending behavior and the punishment. Experimental psychologists recognized this principle decades ago in their attempts to understand how learning influences behavior. So long as we treat physical aggression in an automobile differently from other types of aggressive behavior, the long-term outcome will not be bright.

Simply put, the lack of uniform penalties, the lack of uniform enforcement of regulations, the time delay before punishment is administered, and the relatively light punishment administered all guarantee that aggressive driving will be difficult to control through negative sanctions alone. We must combine these sanctions with other methods of behavior modification if there are to be significant reductions in the cost inflicted on society by aggressive driving.

Mandatory Stress Reduction and Aggressive Driving Classes for Offenders

One of the clearly identified causes of aggressive driving is stress. Stress can be caused by a variety of things, but it is clear that it is frequently mentioned as a cause of aggressive driving. There is some disagreement among professional psychologists who have studied aggressive driving about the potential merits of treating individuals with stress reduction therapy, but learning how to manage stress might go a long way toward improving driving behavior. Individual stress management therapy is expensive and requires the individual to recognize that he or she has a behavioral problem. One possible incentive might be to give drivers who voluntarily go through "prophylactic" stress management therapy a significant reduction in their insurance rates. Insurance providers have long recognized the value of smoke detectors in houses, so why not preventive driving measures? Yet we must acknowledge that changing behavior is difficult for most and may be impossible for the most egregious offenders. Remember the joke about how many psychotherapists it takes to change a lightbulb? Only one, but the lightbulb has got to want to change.

Stress management should be part of the health services that are available to all people. Driver training classes, in particular, should include a large behavioral component highlighting techniques for managing stress, not simply instruction on how to drive. Reducing emphasis on competition in the United States will be difficult because competition is entrenched in Western society. A little competition is good, but too much of it, or any of it in the wrong place, is bad. Signaling submission and the admission of an error is like saying "I am sorry," but there is no way to say you are sorry in a car. Thus frustration builds while driving.[79]

Although it is not a guarantee of rehabilitation, there may be some very real gains in requiring stress management courses for all convicted aggressive drivers. There are a number of techniques that can be taught to help relieve the stresses of driving. Increased awareness of things that are likely to set the stage for aggressive driving (unrealistic deadlines, fatigue, lack of concentration) may reduce some of the stress. Clearly, this technique will not reach all aggressive drivers but, if properly done, could be effective.

Teaching Evolutionary Psychology of Driving Behavior in Driver's Education

Over the last two decades or so, there has been a dramatic downturn in the number of high school students taking driver's education courses. In 1978, a government study conducted in DeKalb County, Georgia, found no reduction in crashes or traffic violations by students who took a driver's education course compared to those who did not.[80] Rather than use these results to redesign driver's education programs or even to replicate the study in another urban area, the federal government gave up on subsidizing driver's education programs in favor of supporting seat belt use and anti–drunk-driving programs. Today, only 40 percent of new drivers complete a formal driver-training course, and that may be a major contributing factor to the 20 to 30 percent failure rate of applicants on their initial driver's license test.[81]

Education, not only about the technical aspects of skid control, parallel parking, and the workings of an internal combustion engine, but also about some basic aspects of human behavior, might be of help. Clearly, an understanding of the mechanics of driving is important, but additional work on the psychological dangers of driving should also be a part of the curriculum.

Changes in Societal Values

Aggressive driving has been around for a long time. But why don't people engage in aggressive walking? People who walk are subject to many of the same societal pressures as drivers, but you do not see pedestrians pushing each other off the sidewalks. Sidewalk rage is not a problem. In crowded, densely populated areas plenty of people walk, but has anyone been shot for not getting out of someone's way on a sidewalk in New York City? Well maybe, but it was most likely because of reasons other than their walking speed or which side of the sidewalk they were on. Even if someone bumps into you while walking, a simple apology is usually sufficient. And even if there is no apology for the transgression, most people overlook it. It is ironic that the same people who are mindful of other pedestrians while out walking undergo a dramatic change when they get behind the wheel.

One possible explanation is that people who walk have the opportunity for face-to-face communication. A simple "excuse me" is often enough to ease the tension of close crowding. In automobiles we do not have the opportunity for submissive gestures that mediate interpersonal conflict. There is no way to say, "I give up, and you win." In a car, there is a degree of anonymity and personal distance that does not tap into those deeply rooted rules governing behavior.

One of the ways to overcome the inappropriate expression of aggression while driving is appealing but virtually impossible, and that is a transformation of our national character. Some argue that remaking our rough-and-tumble, highly individualistic, and competitive culture into one in which everyone recycles kitty litter, drinks latte lite, listens to Enya, and eats high-fiber cereal but no red meat is not all that desirable. For all its dangers, road rage may simply be a corruption of those qualities that Americans have traditionally, and rightly, admired: tenacity, energy, competitiveness, hustle. Something, in other words, to be contained and harnessed by etiquette and social censure rather than eradicated outright.[82]

The competing explanations for road rage and how to minimize it are likely not all wrong. More than one of the alternative solutions may alleviate the problem. The question is whether to try to put a Band-Aid on the problem or try to solve it. Some immediate ways of solving the problem have been discussed: better enforcement, individual behavior modification (such as reduced insurance premiums for people who go through stress management), mandatory stress management for those convicted of aggressive driving, admission of the current state of affairs, banning handguns, and better education programs. But the long-term solution is to recognize discordance in our everyday life and our own physiology. Stress-hormone responses are shaped by years of evolutionary history and are not equipped to deal with the pace of life to which we are constantly subjected. The key may be to change some of the fundamental patterns of behavior in Western society.[83]

3 | Beauty, Blepharoplasty, Barbie, and Miss America

> It is better to be beautiful than to be good. But . . . it is better to be good than ugly.
>
> —Oscar Wilde, *The Picture of Dorian Gray*, 1891

Women, in general, are under more pressure to conform to an ideal standard of beauty than are men. This is because, in part, women quickly learn that their social and economic as well as their reproductive opportunities are directly influenced by their beauty. Physical appearance is not unimportant for men, but in the calculus of female mate choice, male looks are relegated to secondary status when compared to the resources a male can control. From the perspective of evolutionary biology, this is not surprising. In fact, it is entirely predictable.

BEAUTY: AN EVOLUTIONARY PERSPECTIVE

Darwin recognized that there are different constraints on reproduction for what he called the courted and the courting sex among sexually reproducing species. He correctly noted that these roles are not associated invariably with one sex and, depending on local ecological circumstances and phylogenetic history, the roles can be reversed. For most mammals, it is males that are the courting sex and females the courted sex. This means generally that males compete with other males for access to fertile females, and females choose a mate from among the most attractive males.[1]

For us, this typical mammalian pattern of reproductive constraints (male competition and female choice) is somewhat more elaborate. While it is true that males compete for access to females, there is a significant amount of competition among females as well. Given that male quality is highly variable (in terms of their competitive abilities as well as the resources they control) and a female's reproductive success is constrained by the quality of the male with whom she mates, it makes sense that females compete for access to the highest-quality male available.

This is particularly true for humans for two important reasons. First, as compared to other mammals, humans have a long period of infant dependency. While there are good evolutionary reasons why we evolved this long period of dependency, they are beyond our discussion here. This pattern of relatively prolonged infant development is expensive and places particularly stringent demands on mothers. Second, long-term associations between particular males and females can make a significant difference toward offsetting the costs of prolonged infant dependency. While unambiguous evidence for monogamy in our distant evolutionary past is unlikely to be found, it is likely that significant advantages would have accrued to a female and her offspring if a male remained in close proximity. Even if males did not provide food, their proximity might have acted as a deterrent against predators. If they also obtained food and shared it with a female and dependent offspring, the selective advantage for male parental investment could have been considerably greater. For humans, the old saying that a good man is hard to find may have particular significance. Unlike competition among human males, female competition is largely based not on physical prowess, strength, endurance, or hunting ability but rather on beauty and the ability to attract and keep male partners.

Male preference for youth in a mate is an obvious example of a preference for women with high reproductive value.[2] Because reproductive value is related intimately to age in females, this means that for humans there is strong selective pressure for males to favor females with certain phenotypic characteristics that are reliable indicators of age. The situation for males is a bit different. Male reproductive value is tied to age, but in another way. With increasing age, many males are able to accrue more and more resources and consequently increase their attractiveness to females. (The arena of mate choice is a complicated one and many other factors play a role, but for this discussion it is helpful to explore some of the areas where our culture and our biology run in opposite directions.)

Beauty may be in the eye of the beholder, but our eyes and minds have been shaped by millions of years of evolutionary history. Our ancestors had access to two kinds of evidence about a woman's reproductive value: (1) her physical appearance, including for example clear skin, good teeth, bright eyes, lustrous hair, good muscle tone, appropriate body fat distribution; and (2) her behavior, such as a symmetrical gait or animated facial expression. These data spoke directly to issues of health and youth and indirectly to reproductive potential and have been hypothesized as key ingredients of male standards of female beauty.[3] If this hypothesis is correct, ancestral males evolved a preference for females with these cues. Men who mated with females with the highest reproductive value left the most offspring, and, to the extent that these preferences had some genetic basis, the genes influencing such preferences would be passed on to their male offspring.

Male preference for signs of youth and health has been documented by anthropologists in a variety of cultures. Among the Trobriand islanders, Bronislaw Malinowski, a Polish-born British social anthropologist, noted in the early twentieth century that the essential conditions of beauty were signs of health, including normal body build, absence of mental or functional disorders, strong growth of hair, good teeth, and smooth skin.[4] It is not likely that any culture would favor signs of disease over signs of health, and it is virtually universal among cultures studied that individuals suffering from leprosy, yaws, ringworm, or tertiary syphilis are shunned and deemed unattractive.[5] In some societies obesity is seen as a desirable characteristic. This is particularly true among certain West African groups, where women are placed in a fattening hut before marriage and gorge themselves in order to gain as much weight as possible.[6] Among the Annang of Nigeria, prior to marriage young women are sequestered in what is called a "fattening room." Parents may put their daughters into the hut or prospective husbands put their fiancées there in order to increase their desirability. During her period of incarceration (lasting from three weeks to seven years), a woman was allowed to do no physical work. She was bathed daily, and her skin was rubbed with oil to make it soft and gleaming. She was fed three meals a day and was expected to eat all the food presented. She was also instructed in the womanly arts of pleasing her husband as well as her mother-in-law.[7] The underlying basis for this aesthetic preference may be the reported positive association between fat and childbearing.[8]

Some experimental evidence suggests how pervasive is the preference for youth as a criterion for female beauty. Men and women were given a series of photographs of women of different ages and were asked to rate their attractiveness. Regardless of the age or sex of the rater, judgments of facial attractiveness declined with the increasing age of the women photographed. Interestingly, the attractiveness ratings declined more rapidly for the male raters.[9] This finding is consistent with the evolutionary importance of age as a cue to males of a woman's reproductive value.

There is considerable debate over how the individual develops particular preferences for beauty. According to traditional psychological dogma, standards of attractiveness are learned through cultural transmission and therefore do not emerge until children are three to four years old.[10]

However, other researchers found that infants two to three months of age as well as infants six to eight months old, when shown slides of adults of varying degrees of attractiveness, gazed longer at the faces that were rated attractive by independent evaluators.[11] One-year-olds also showed more play involvement, less distress, and less withdrawal when they were interacting with strangers wearing attractive masks than when they were exposed to experimenters in unattractive masks. Researchers also found that one-year-olds played longer with dolls independently rated facially attractive than with those rated unattractive. These data suggest that preferences for certain facial features are expressed early in life and are not the outcome of a long process of enculturation.

So, basic standards of beauty seem to develop early in life and seem to be largely independent of early experience. True enough, the specific details are learned more slowly and vary cross-culturally, but the basic components of attractiveness vary little across cultures. Research psychologists asked people of different races to rate the attractiveness of Asian, Hispanic, black, and white women in a series of photographs. Researchers found a statistically highly significant correlation among all of the raters.[12] This means that in this study, regardless of racial or ethnic identification of the rater, individuals consistently ranked the same women as attractive.

The obvious question that arises from these observations is, what makes a face attractive? A growing body of empirical data suggests that the fundamental key to attractiveness is not some particular characteristic but a more general quality, that of symmetry.[13] In a variety of experiments, it has been demonstrated that people carefully attend to the symmetry of the face and the body, and they consistently choose individuals that have the most symmetrical features as the most attractive.[14] This has been demonstrated not only with photographs of individuals but also with computer-generated photographs in which the features could be experimentally manipulated. Researchers measured a variety of human phenotypic characteristics and found that those subjects who exhibited quantitative asymmetries were judged less attractive by an independent set of raters.[15] Given that the basic selection criteria seem to be universal for humans, it is likely that some aspects of the selection criteria have a Darwinian basis.

These results all suggest that symmetrical features are important in evolutionary terms, but what are the causes of asymmetries? A wide variety of environmental insults can lead to asymmetries, as well as irregularities in development both pre- and postnatally. In addition, the normal processes of aging produce asymmetries, so it is likely that those individuals with the most symmetrical features tend to have experienced normal sequelae of development, tend to be young, and tend to be positively associated with other indicators of physiological and psychological well-being.[16] The degree of symmetry in phenotypic features augurs well as a proxy measure for overall reproductive fitness, and consequently the preference for mates with symmetrical features could have been highly selected during our evolutionary history.[17]

IDEALS OF BEAUTY, CULTURAL DEFINITIONS, AND MASS MEDIA

While there is little consensus on the *specific* characteristics (fat vs. thin, light skin vs. dark skin, etc.) considered beautiful in a cross-cultural sense, there is considerable agreement on the importance of symmetry and other indicators of good health as more *general* characteristics that define beauty. Rather than some specific characteristic, it is likely that all cultures favor signs of good health in both sexes, such as smooth skin, good teeth, symmetric features, a uniform and symmetric gait, but also firm breasts and hips, roundness rather than angularity, and fleshiness rather than flabbiness in women.[18] It is important to recognize that considerable variation in specific standards of beauty is completely consistent with more general evolutionary indicators of high reproductive value. Clearly, culture plays an important role in defining specific criteria against the larger evolved background of mate-choice characteristics. Standards for men are likely to be specified in ways similar to those for women in terms of overall health and well-being but are likely also to include evidence of resources.[19]

One finding of those who have studied standards of beauty is that specific standards change over time. One of the main reasons for rapid changes in concepts of beauty is media coverage. Evidence for this can be found in a historical look at media in the United States. Until the widespread use of the radio,[20] newspapers were the source of information for the masses. One of the problems that early newspapers experienced was a lack of advertising, but that quickly changed. In the late 1920s, the radio, coupled with the newspapers, ushered in a whole new era in communication. Commercial advertising was born, and today companies spend hundreds of millions of dollars annually in efforts to persuade consumers to purchase their products. The average American is bombarded by 1,500 to 3,000 marketing messages per day. Now this may seem like an extraordinary number, but imagine that these messages come in a vast array of forms ranging from radio jingles and billboards to apparel logos, Internet banner ads, free samples left hanging on doorknobs, plus the obvious newspaper, magazine, and television advertising. In fact, the average American will spend over a year and a half of his or her life watching commercials on television.[21] Americans are admonished to buy "Brand X" because it will make them thinner, healthier, richer, sexier, smell better or not smell at all, or it will make life easier, more fun, less painful.

Research on the effects of the media in establishing preferences for particular products among children is quite clear. In a study of nine- to ten-year-old girls exposed to commercials for lipstick and diet soda, researchers found that subjects intended to use the products to which they were exposed when they became adults.[22] While this does not conclusively establish a precise link between early preferences and later purchases, it does serve to emphasize that early exposure is an important part of successful advertising. Not only does early exposure to certain products influence later consumerism, it also influences one's behavior, attitudes, and extent of social integration into peer groups.[23]

Why are we so easily influenced by advertising? Entrepreneurs realized generations ago that it is easy to sell items that touch our basic biology. There is always a ready market for a product or device that will cure what ails us, enhance our sex lives, make us feel better, or vanquish signs of aging. Our vulnerability to advertising practices that target any aspect of sexuality is particularly apparent. If we can be convinced that a product will make us more attractive to the opposite sex, then it is likely to succeed, since attracting mates is a part of our evolved psychological predisposition.

So, in a sense, we are gullible. Why do we so readily accept the claims presented? We do so because it is part of our evolutionary heritage. Of course, I don't mean that there is a gene for gullibility. But among our ancestors, those who readily adopted some novel practice, decoration, or attribute were better able to distinguish themselves from the competition. The tendency to readily accept such new ideas is particularly evident when it comes to attracting the opposite sex. We know that novelty is a powerful stimulus to sexual behavior in a variety of different species including monkeys, mice, domestic fowl, and humans.[24] It should come as no surprise then that our quest for novelty is a part of our evolutionary heritage.

One of the by-products of our willingness to accept novel ideas is that sometimes others will take advantage of the situation. Exploitation of our gullible psychological nature must have been figured out early in our evolutionary past. In a sense, we are responsible for the unbridled success of those selling cure-alls. Deception surely must have evolved long ago, and certainly it is an intrinsic part of human sociality. And, unfortunately, it is also a part of some unscrupulous advertising. So the question is, why are we

susceptible to deceptive advertising practices? It is likely that humans developed a certain degree of gullibility as an adaptation to social living.[25] We
are often not very good at separating fact from fiction and thus leave ourselves open to a variety of outside influences. As evidence of this tendency,
one can ask almost any woman and quite a few men about the number of
cosmetic products they have bought and tried that produced no discernible
results. Why do we fall for the same sales pitch over and over, particularly
when it comes to attracting the opposite sex?

The point is that the advertising industry has developed highly sophisticated ways to influence our behavior on a variety of levels. One of the targets for the health and beauty industry is to define culturally accepted
standards of beauty for both sexes. If industry-influenced standards are
adopted by the majority of people in a particular group, then the industry
can shape preferences in economically favorable ways. One particularly effective area of advertising is fashion.[26] Exploiting our gullibility and strong
desire to maintain group identification lies at the heart of the success of the
fashion and beauty industries.

Ideals That Come and Go

While fashion certainly plays a role in defining beauty, the real definition
comes from artists and illustrators.[27] There is a widely held notion that preferences for puny or plump were widely accepted in the past, with cultures
favoring either one or the other. This is simply not the case. Take the Renaissance, for instance. The full-figured, some would even say corpulent,
women painted by Titian (1477–1576) were painted at the same time as the
svelte female subjects of Lucas Cranach the Elder (1472–1553). The nudes
of Diego Velásquez (1599–1660) are modern in comparison to the rotund
nymphs of Peter Paul Rubens (1577–1640), while both artists were working at the same time, although in different cultures.[28] Clearly, the characteristics of specific definitions of beauty truly are in the eye of the beholder
but are also strongly influenced by basic cross-cultural preferences.

Beauty in the United States

In order to understand the role that cosmetic surgery plays today in helping to define what is beautiful, it is important to take a brief diversion into
the history of beauty in the United States. Two distinctly different types of
beauty arose in the United States in the nineteenth century and were represented in a wide array of media. On one hand were the images captured

in the famous Currier & Ives lithographs. The women in these works are best characterized as slight and delicate with small hands and feet, white skin, and just a blush of color in their cheeks. This image of beauty is a portrait of frailty and fragility. Proper behavior and high social status are hallmarks of this type of beauty.

The Currier & Ives woman was challenged in the mid-nineteenth century by another type of woman, the big beauty. Women who were endowed with larger busts, wider hips, and all-around fuller figures rose in popularity in the latter half of the nineteenth century, with women like Lillian Russell commanding the admiration of men across the United States, even in the upper classes. A healthy appetite and padding were the order of the day in the 1880s, but these preferences were destined to pass.

By the end of the century, the full-figured, busty look was a thing of the past, particularly among the upper class. The full-figured ideal did not die a quick death and, even prior to World War I, was still a favorite in burlesque houses and on the pages of the *National Police Gazette*.[29] Interestingly, the first calendar nude appeared in 1913, and, rather than being one of the big beauties, the woman pictured was a much smaller and more delicate individual. The model for *September Morn*, as the image was named, was one of the Currier & Ives models. The painting was done in 1912 by Paul Chabas and created quite a controversy when it was first shown.[30] The woman is rather slender and delicate, with her arms covering her body, in a strikingly modest pose.

One of the early influences on the standards of beauty in the United States was, oddly enough, not a woman at all but an artist and illustrator for *Life* magazine, Charles Dana Gibson. His Gibson Girl set the standard for a new type of beauty. First appearing in 1890, she took the shape of the fragile Currier & Ives figure, slender with a visible sense of respectability, but she also had the large bust and hips of the opposing full-figured ideals. She piled her hair on top of her head and wore a corset. Occasionally she would appear in sport clothes or swim suits, revealing the previously invisible feminine leg. However, the popularity of the Gibson Girls was short-lived.

In contrast to the Gibson Girls style, postwar fashion took a distinctly different swing. The dresses of the 1920s took on a shapeless, boyish, and almost formless appearance. Unlike the corset, undergarments tended to flatten rather than accentuate the silhouette. The Gibson Girl gave way to the flapper of the Roaring Twenties. A woman's attractiveness was judged not by the shape of her torso but by her face and legs. The female torso became a flattened tube, and the emphasis on bust size was completely reversed, to the extent that women with large breasts would wear "correctors" or "flatteners" in an attempt to conform to prevailing fashion.[31]

It was during this period that the full force of the media on fashion began to be felt. With the advent of inexpensive radios and the widespread circulation of newspapers, the messages of fashion could be transmitted with unprecedented speed. The Sears catalogue was one of the primary fashion arenas for much of America, and people had access to all sorts of fashions by mail order. Flappers would remain in popularity until the infamous stock market crash of 1929 and the resulting economic depression. After that, breasts and curves would return. In the 1930s, hemlines fell and narrow waistlines returned. Unlike the stars of the 1920s, Jean Harlow, Mae West, and Greta Garbo would display fuller figures and the return of an emphasis on female secondary sexual characteristics.

The post-Prohibition feminine ideal was again strongly influenced by the illustrations of one man, George Petty. Petty was a chief illustrator for *Esquire*, and, beginning with the first issue in 1933, the Petty Girl would define a new set of beauty standards. Unlike her predecessor the Gibson Girl, the Petty Girl was a notorious gold digger. She had the long legs of her flapper predecessors, now coupled with stockings and high heels. It is with the Petty Girls that the leg surpassed the breast as the erotic symbol in fashion.

Esquire hired a new illustrator, Albert Vargas, in the early 1940s. While it is certainly true that no single person can define fashion for an entire nation, Vargas did express prevailing values with considerable authority. Like Petty Girls, the muscular women from Vargas's pen showed larger and more rounded breasts than those of his predecessor. From this point onward through the 1960s, the ideal breast size in America increased, due in no small part to the pen of Albert Vargas.[32] During the 1950s, Hollywood played an important role in defining beauty. Large breasts, a small waist, and a seductive walk were the currency of the day. As you might imagine, padded bras, girdles, silk stockings, garter belts, and high heels were in great demand.

In 1953, Hugh Hefner founded a magazine that was nothing more than a sexy imitation of *Esquire*. *Playboy* was aimed at young educated males, and for years defined at least one standard of beauty in the United States. Hefner went so far as to hire Vargas away from *Esquire* as he continued to capture the ideal of feminine beauty in his famous nude illustrations. *Playboy* attempted to define standards of beauty with the introduction of the "Playmate of the Month," centerfold pictures that would become the most popular pinups in the world. Marilyn Monroe was the first Playmate and appeared in a shockingly naked pose for magazines at that time.

In addition to the voluptuous standards of beauty, there were also more slender ideals as well. Grace Kelly and Audrey Hepburn defined a style of high-fashion, classy sensuality and elegance for the 1950s. Marilyn Monroe and Sophia Loren embodied a more blatant, earthy sexuality. At this time there began to be a clear recognition of the differences in these two body types, and while one was featured in pinups, the other was featured in high fashion.

The 1960s were a turbulent time for many aspects of American society, and not the least of these was the standard of feminine beauty. The pulchritudinous images of the 1950s gave way to a progressive and what would turn out to be almost relentless slenderization of the ideal of American women. *Vogue* and *Cosmopolitan* would set the defining standards for beauty in the post-Vietnam era. Along with the Beatles came miniskirts, panty hose, and Leslie Hornby Armstrong (aka Twiggy), all ninety-seven pounds and 31-22-32 of her, to grace the pages of *Seventeen* and *Vogue*.[33] The emaciated look of Twiggy was popular and played an important role in the trend toward more slender figures. Even Playmates became taller and more slender, and the emphasis on breasts gradually shifted again to legs and buttocks.

The introduction of the French bikini is credited by some with initiating the emphasis on thinness in the United States. While European women seemed to be less concerned about their appearance in these minuscule bathing costumes, American attitudes reflected the influence of the Puritans and John Calvin. Corpulent figures and bikinis just did not go together in the United States, with the outcome being yet another period of pressure to diet and be thin. These changes are demonstrated no more clearly than in the physical dimensions of commercial depictions of ideal beauty. In 1894, our ideal woman was five feet four inches tall and weighed 140 pounds. In 1917, the physically perfect woman was five foot four inches tall and weighed 137 pounds. By 1947, she was down to 125 pounds, and by 1975 she was down to 118 pounds despite having increased four inches in height.[34] So twenty-five years ago, models weighed 8 percent less than the average woman, but today models weigh 26 percent less.[35] Today's ideal fashion model should be no less than five feet nine inches tall and weigh no more than 115 pounds, and her age is anywhere from fifteen to nineteen.

The continued quest for thinness is fueled by a number of factors, not the least of which is the widespread adoption of ideals that are beyond the physical capability of most people. Simply put, there are two basic ideal body types that have emerged in recent years. One is based on standards set by the insurance industry for "normal" height/weight ratios. The Metropolitan Life Insurance Company (MetLife) set the standards for ideal and desirable weights, defined as those with the lowest mortality, for the first time in 1940. More recently MetLife has made an effort to take body frame size into account, which has slightly raised the weight for height values. These data are obtained from population-wide surveys and are expressed as a mean and a measure of statistical variation. For lack of a better term, these may be called the *biological* ideals. Individuals who conform to these ideals have been shown to have the lowest mortality. While these ideals ignore quality of life issues, they certainly are useful guidelines.

On the other hand, there are what we might call *cultural* ideals. These ideals are effectively decoupled from any biological reality and seem to be much more extreme in their standards. The existence of two standards is not necessarily a problem so long as people understand the differences and behave accordingly. Unfortunately, for a proportion of the population, cultural ideals take precedence over biological ideals, resulting in complicated and confusing behaviors.

Body Image Dissatisfaction

The net result of the discordance between the cultural ideal body type and the biological ideal body type is a widespread dissatisfaction with physical appearance among both men and women in the United States. In reality, most Americans have considerable difficulty achieving the biological ideal body weight, much less the cultural ideal. This incongruity sits at the heart of much of the psychopathological behavior associated with body image dissatisfaction. While there are many dimensions of body-type dissatisfaction, two well-known dimensions serve to illustrate my point. They are weight and skin color.

Weight

The emphasis on thinness as a defining characteristic of beauty signaled the beginning of the diet craze in the United States. While large breasts were in vogue, dieting was not all that necessary, but as soon as more slender hips and legs became the currency of attractiveness, diets reasserted themselves. The efflorescence of the diet and exercise industry in the 1970s is well known and has continued to develop through the twentieth century into the twenty-first. Given the enormous influence of mass media on our cultural ideal body type, it is no wonder that American women have been in hot pursuit of slender bodies. One of the consequences of socialization into the culture of slenderness is the enormous dissatisfaction that many women and even young girls feel toward their bodies, and in particular their weight.

In order to overcome this rampant concern with body weight and fatness in the United States, there has developed a multibillion dollar health, fitness, and dieting industry. In 2000, Americans spent about $30 billion on diet programs, diet foods, and weight loss plans, nearly $2 billion on running and exercise shoes, $3.6 billion on fitness and exercise equipment, and $7.5 billion on cosmetic surgery, all in the hopes of looking a few years younger, a few pounds thinner, sexier, and generally more physically attractive.[36]

Skin Color

In addition to height and weight, numerous other body characteristics have received attention in the quest for perfection. For many centuries in Japan, India, and much of the West, there has been a clearly understood association between skin color and social status. Even in Egyptian society at the time of Cleopatra,[37] powder and other makeup was used to lighten skin tones. The preference for light-colored skin among light-skinned populations has persisted throughout much of human history. In Victorian England, light skin was venerated, and considerable effort was expended to keep women, particularly the upper class, out of the sun.[38] This preference for light skin color made its way to the United States, and by the end of the nineteenth century a variety of skin crèmes had been developed specifically to lighten the skin. The association of light skin color and the leisure class had been established.

By the 1920s, attitudes had changed. Paleness was out, and suntanning had become a craze, due at least in part to the popularity of a much photographed cinnamon-toned Coco Chanel cruising to Cannes on the Duke of Westminster's yacht.[39] The widespread preference for darker skin challenged one of the basic assumptions of the cosmetic industry—that good skin was light skin. A few companies produced "sunburn" tints in the early 1920s, but it was generally in the last part of the decade that tanning lotions and darker face powders were generally available.

Advertisements touted the health benefits of dark skin, but the implied message was that there should be a clear separation between the races.[40] In the 1930s and 1940s, the medical profession began to extol the virtues of sunbathing as particularly beneficial for children suffering from *Lupus vulgaris*.[41] In recent years, the preference in Western society among whites, and particularly among middle- and upper-class individuals, has been away from light, pale skin to a year-round tan.[42] This preference was developed partially as a consequence of greater amounts of leisure time, as well as the invention of the sunlamp and, more recently, tanning salons.[43] In 1996, Americans spent almost $400 million on suntan lotions and other cosmetics to make them tan more rapidly and more evenly, sometimes even without the sun.[44]

Most suntan worshipers, however, shun the "fake tan" cosmetics and favor the real thing, exposing themselves to sublethal whole-body UV radiation. Yet artificial tans have had some popularity, particularly in the 1950s, when John F. Kennedy used Man-Tan, a self-tanning makeup, before his television debate with Richard Nixon.[45] There are a number of well-known and scientifically documented deleterious consequences of sun exposure, including skin cancer, premature aging, and drying and wrinkling of the skin.[46] But, in spite of widespread knowledge of the risks of exposure to the sun, the rise in the popularity of suntanning is paralleled only by a rise in the risk of developing malignant melanomas, with the frequency increasing from 1 person in 1,500 in 1930, to 1 in 250 in 1981, to 1 in 87 in 1996, to 1 in 79 in 1999, and this risk is likely to continue increasing at about 7 percent per year.[47] The perceived benefits of a suntan clearly continue to be widely held in Western society, not only by adults but also by adolescents and children, who equate a suntan with being healthy and attractive.[48] Even though people are aware of the potential dangers of exposure to the sun, in the eyes of many, the benefits of a tan outweigh the costs.[49]

Culture of Dissatisfaction

A number of studies have suggested that women are more concerned with appearance than are men. This is not to say that men are not concerned about their appearance, given the amount of resources that men invest in clothes, hairstyles, cosmetics, and so on. But it has been argued that appearance is of greater importance to women than men in terms of interpersonal relations, professional advancement, and overall sense of well-being. Physically attractive women are reported to have greater social power, self-esteem, and positive responses from others.[50]

In 1927, Hazel Rawson Cades, longtime beauty editor of *Women's Home Companion*, published a beauty guide for young women titled *Any Girl Can Be Good Looking*. She argued that "everybody is thinking these days about good looks. The bar has been raised for the jump. The passing mark is higher. Being good looking is no longer optional. . . . There is no place for women who are not."[51]

Women are told from an early age that, if they are diligent in their efforts, there is nothing they cannot change about themselves. Those who ignore their potential are seen in a very negative light and are described as lazy, weak-willed, and slovenly. Women are socialized into a pathological level of dissatisfaction with their appearance and an unrealistic view of their ability to alter it.

Researchers have argued that repeated exposure to attractive models in advertising and other media dictates, to a large extent, an individual's idea of what is an acceptable physical appearance.[52] The ideal standard set by fashion models and TV personalities is not only unrealistic for the average person, it is unattainable.[53] Moreover, comparison to this unattainable standard leads to increased dissatisfaction with one's own body image in a number of populations.[54]

Interestingly, children are brought into this culture of dissatisfaction early in life. The effects on children of TV ads for adult products have been well studied, and it has been convincingly demonstrated that children perceive these advertised products and brands as associated with being an adult and thus as products they should consume. Even if children are not consumers of a particular product, they can be influenced to buy that product in the future. One has to wonder how the advertising of products today that promise to shape, firm, reduce, enhance, color, and otherwise alter parts of our bodies will affect perceptions of today's children when they grow up. Perceptions of ideal body types are likely to be as easily influenced as preferences of which beer to buy. If that is true, then advertising has played an enormous role in selling the American public a bill of goods about the desirability of a virtually unattainable body shape. Research has shown that, by age six, children acquire prevailing cultural values of beauty, and thinness is accepted as a desirable physical attribute considerably before puberty.[55] This means that preferences and attitudes are shaped much earlier than many parents may imagine.

Barbie Dolls, or Don't Trust Anyone Who Tries to Sell You a Doll That Can't Stand Up

The Barbie doll is one of the most popular toys ever created. Today Barbie is sold in 140 countries at the rate of two dolls every second. Barbie was developed initially by Ruth Handler, the daughter of immigrant Polish-Jewish parents (née Moskowicz), who along with her husband, Izzy Handler, founded the Mattel Toy Company. Barbie was first marketed in 1959. In 1997, the annual sales of the Barbie Division of Mattel were over $ 1.9 billion.[56] Since Barbie's introduction, over one billion Barbie dolls have been sold.[57] In fact, the average white middle-class American girl between eight and ten buys three Barbies a year and has eight in her collection, compared to one in the early 1980s.[58]

Researchers have investigated which characteristics of Barbie have contributed to this phenomenal success. Long legs, long necks, square shoulders (vs. sloping) angled in, uniform teeth, long shins, strong chins, large eyes, knock-knees (vs. bowlegged) and plantar (vs. dorsal) foot flexion were shown to be preferred strongly in a survey of nearly five hundred individuals.[59] Interestingly, if one examines the typical Barbie, she appears to be tall, long-legged, slim, almost wasp-waisted, with a long neck, red lips, large eyes, and square shoulders. She has a short torso, straight evenly spaced teeth, smooth and hairless skin, plantar foot flexion (walking on the balls of her feet), long straight fingers, a nonsloping forehead, flat abdomen, and is not bowlegged.[60]

Since Barbie and her partner Ken are marketed as "aspirational models" by their manufacturer, it is a fascinating exercise to scale the dolls to adult size.[61] To attain Barbie dimensions, a five feet two inch, 125-pound female with 35-28-36 measurements would have to gain twenty-four inches in height, increase five inches in the bust, and decrease six inches in the waist. To put it a bit differently, our ideal would be seven feet two inches, with measurements of 40-22-36.[62] Interviews of junior and senior high school girls found that their perception of the ideal girl has at least a modest resemblance to Barbie: she is five feet six inches tall, weighs 110 pounds, wears a size five, and is able to eat whatever she wants and never gain weight. This does not sound like a cultural ideal to me but rather like the making of a young woman with an eating disorder.[63] Of course, the real question is the extent to which little girls perceive fashion dolls, such as Barbie, as an ideal to be achieved. The answer is that we do not know for sure, but considerable evidence suggests that such exposure creates expectations that are simply beyond the physical capabilities of most individuals.

While Barbie's popularity may be worldwide, there are some non-Western societies that would like to see this part of Americana return home. In Malaysia, the Penang Consumer's Association has called for Barbie dolls to be banned, citing the negative effects of the dolls on children. The dolls' blonde, leggy, and non-Asian features were cited as characteristics that did not contribute to a positive self-image for Asian children.

Opposition to Barbie can also be found in the United States. The Barbie Liberation Organization (BLO), a group of San Diego art students, swapped the voice box in the Talking Barbie with the voice box for Talking

GI Joe, with some interesting consequences. Barbie has also "come out" in a homosexual bar to unite with her true love, the transsexual Kendra, in a chapter "Barbie Comes Out" in the book *Mondo Barbie*. In addition, *Mondo Barbie* features a variety of other revealing tidbits about Barbie, such as "Hell's Angel Barbie," "Twelve-step Barbie," "Barbie Meets the Scariest Fatso Yet," and "The Black Lace Panties Triangle."[64] Variations of Barbie have appeared as David Goldin's "Temptations of Barbie," in which the doll is surrounded by sweets, and in Douglas Fraser's "Biker Barbie," straddling a chopper.

TEMPORARY BODY ALTERATION AND THE QUEST FOR BEAUTY

Cosmetics

The quest for physical beauty is an ancient human pursuit. There is considerable evidence that ancient Egyptians used cosmetics as early as twelve thousand years ago, protecting their bodies with colored earths and clays along with grease and with oil from the castor plant.[65] The earliest historical record of cosmetics comes from the First Dynasty of Egypt (ca. 3100–2907 B.C.). Tombs of this era have yielded unguent jars, and, from remains of later periods, it is evident that the unguents were scented. Such preparations, as well as perfumed oils, were used extensively by both men and women to keep the skin supple and unwrinkled in the dry heat of Egypt.

Ovid advised Roman women to use natural products to improve skin tone and quality, to wear black if one's hair was blonde, and to use rouge if there was a lack of color in the cheeks. Women of the upper classes in Rome used a special soap imported from Gaul and bleached their hair to a golden blonde. Cleopatra painted her upper eyelids blue and her lower eyelids green. She used kohl to draw in the shape of her eyelids and to accentuate her eyebrows, while she used ceruse to whiten her face, neck, and breasts. Yellow ocher was used for her cheeks and carmine dyes for her lips. All in all, Cleopatra redefined what it meant to be "made up."

By the middle of the first century A.D., cosmetics were widely used by the Romans, who employed kohl for darkening eyelashes and eyelids, chalk for whitening the complexion, rouge and depilatories, and pumice for cleaning the teeth. In ancient Greece, a number of substances were used to adorn the face: kohl was used to darken the eyes, and ceruse for whitening the face.[66] Ancient Greeks often did not use these products with great skill, as was noted by the Greek satirist Lucian: "If one could see these painted women getting out of bed one would find them even less attractive than monkeys."

The Crusaders found cosmetics widely used in the Middle East and began to spread their use throughout Europe. Yet their use did not become widespread until the twelfth century or so. Traders and explorers began to bring back strange Eastern substances that included perfumed oils, eau de Chypre, dyes for lips and eyes, and a white paste for the skin. By the middle of the twelfth century, fashionable young women in England were shaving the hair from the front of their heads as well as from their eyebrows. This gave them the appearance of having a very high forehead. The remaining hair was often dyed bright crocus yellow; faces were painted with a white paste and the lips stained a deep red.

In the time of Elizabeth I, a variety of products were used to color the skin, but one of the most important was the substance used by Cleopatra. Ceruse consisted of lead oxide, lead hydroxide, and lead carbonate. White lead was first mixed on a palette with egg white and sometimes vinegar and was applied to the skin with a damp cloth. Sometimes it was overlaid with varnish that, when dried, must have dramatically restricted the mobility of the face. Kohl was used for darkening eyelids, lashes, and brows. Malachite was crushed and used as a green eye shadow. Red fucus, a strong dye made of mercuric sulfide, was used to color lips. Freckles were removed with sublimate of mercury. Sulfuric acid, turmeric, and alum water were used to remove unwanted hair. The regular use of all these preparations, many of which are toxic, must have had drastic health consequences.[67]

Face painting continued to have widespread popularity in England, so much so that in 1770 Parliament attempted to enact penalties similar to those for witchcraft against women who used cosmetics or artificial aids to entrap men into marriage:

> . . . whatever age, rank, profession or degree, whether virgins, maids or widows that shall impose upon, seduce and betray into matrimony any of His Majesty's subjects by the scents, paints, cosmetic washes, artificial teeth, false hair, Spanish wool, iron stays, hoops, high-heeled shoes, bolstered hips, upon convictions the marriage shall be null and void."[68]

Cosmetics in England decreased in popularity under Queen Victoria, and, even after her reign, they returned only under the guise of healers. Prior to World War I in the United States, cosmetics were expensive and largely confined to the affluent. After World War I, cosmetics became cheaper, and a generation of emancipated women from the ammunition factories and other wartime industries were able to purchase them. However, overuse and unskilled application quickly became a sign of the working class. Today, there is a complicated procedure associated with the application of the various types of crèmes, salves, ointments, gels, pastes, shadows, and shades used in cosmetics. The current widespread use of cosmetics and the economic success of companies manufacturing the products is due in no small part to their appeal to the narcissistic aspects of modern society. The message is that, with just the right combination of cosmetics, all facial imperfections and perceived inadequacies will disappear magically.

Let me not appear overly critical of women in their use of cosmetics, for men have certainly joined the movement. Products designed to slow or mask the aging process, improve one's smell, lighten one's teeth, and regrow hair have gained considerable popularity in recent decades among men. There has also emerged a male counterpart to *Cosmopolitan*, called *Men's Health*. Since 1986, the year of its introduction, sales for *Men's Health* have reached 1.3 million monthly. Male grooming products have reached over $3.3 billion in sales in the United States, and estimates predict growth at over 4 percent per year. In addition, nonsurgical hair replacement is rapidly growing in popularity. With an estimated forty million bald men in the United States, this is a salesman's dream. Pharmacia & Upjohn, makers of Rogaine™, estimate that men are spending over $65 million per year on minoxidil (the active ingredient in Rogaine) products, $400 million on wigs and toupees, and another $100 million on what some people might call quack remedies (e.g., elixirs, teas, and horse-hoof ointments). Sales of hair dyes for men have tripled in the last decade, and it is estimated that today men spend about $100 million annually to hide the gray.[69]

Now what do wearing makeup, hiding the gray, and being tan have to do with biology? Well, this is one of those interesting cases of our culture and our biology seeming to operate in parallel fashion. Clearly, attracting mates is an exceedingly important part of our evolutionary history, and the use of artificial aids is consistent with that overall goal. The only thing that might be seen as a bit odd is the extent to which the culture of makeup has evolved, but given the affluent, narcissistic culture we find ourselves in, it ought not be all that surprising. Nor should we be surprised to find that men as well as women are vitally concerned with improving their appearance.

Hair Styles

Of course there are a variety of additional techniques that are used to enhance appearance, and skillful manipulation of the hair around the face has long been recognized as an important part of the whole package. Today, both men and women squander lavish attention on their hair in efforts to change its color, to hide any signs of graying, to curl it if it is straight, to straighten it if it is curly, to make it silkier, softer, and more manageable, or to make changes just for the sake of novelty (more about the importance of novelty in the evolutionary game later). It is as if one's hair is the enemy and must be brought into line, and we employ a legion of specialists to help us control those unruly locks.

Beauty culture was in full bloom in the United States in the early twentieth century, and the idea that beauty was something that could be acquired if one had sufficient resources was widely accepted. *Woman Beautiful* magazine reported in 1907 that there were 36,000 female hairdressers, 25,000 manicurists, and 30,000 specialists in face massage and skin culture. The United States Bureau of Labor Statistics reports that in 1998 there were 605,000 employed hairdressers, hairstylists, and cosmetologists, along with 54,000 barbers, 49,000 manicurists, and 15,000 shampooers.[70] This is a considerable workforce considering there were 681,000 lawyers in 1998, according to the same *Labor Statistics* report.

Now don't misunderstand; this is not simply an aspect of Western society. Men and women around the world color, clip, curl, iron, pull, crimp, plait, paint, and style their hair. Anthropologists have spent much of the last hundred or so years documenting how people in cultures unlike our own solve the problems of everyday life confronted by all humans. One of the many things that ethnographers are concerned with is the material culture of a society, and one dimension of that material culture is hairstyle. If we look cross-culturally, hair adornment is not restricted to women. One of the most interesting hairstyles and one that is well known to anyone who has traveled in East Africa is the hairstyle of Masai warriors (*moran*). Young warriors let the hair grow, and, when it has the necessary length, they twist it between two fingers into interlaced spirals. Somewhat later it is twisted together into strands and coated with a mixture of animal fat and red clay. The hair is lengthened further by twisting in fibers from the bark of the baobab tree. Hair from the back of the head, lengthened up to twenty inches, is placed around a foot-long stick, and on this it is then wrapped tightly onto goat leather so that it forms a stiff braid reaching to the waist. Over this braid one frequently finds one to three smaller ones. Often the temple braids are joined with the forehead braid and all are tied together under the chin.[71] Now by any standards this is an elaborate hairstyle and one restricted in its distribution even within Masai society.

In order to understand more fully the place that hair has in the quest for beauty, it is worthwhile to take a brief diversion into the history of hair and hairstyles. Hairdressing was an important aspect of life in early Egyptian society, at least for the affluent. Hairdressers were depicted prominently on frescoes, urns, and ceremonial coffins. Shaved heads and smooth, hairless bodies were signs of Egyptian nobility by about 3000 B.C., but fashion required men and women to wear wigs of real hair or sheep's wool. False beards were popular along with wigs dyed using indigo to achieve the favored black color. Henna, a powder made from the leaves of a shrub (*Lawsonia inermis*), gave hair, nails, and toes a red-orange cast. Somewhat later (after 1150 B.C.), wigs were dyed different colors such as red, green, and blue.

The first styling salons came into existence with the ancient Greeks, where customers were shaved, curled, manicured, pedicured, and massaged. Some barbers were skilled artists and respected community members. Others were household slaves who were punished if they allowed a hair out of place. Hair was mainly thick and dark and worn long and naturally curled. Blond hair was rare and admired by the Greeks, and both men and women bleached their hair with potash water and infusions of yellow flowers. By the fifth century B.C., Athenian men began to wear shorter hair, while beards remained popular for another hundred or so years. The demise of beards, at least in the military, was for tactical reasons. Soldiers of Alexander the Great had to shave their beards to avoid having them seized in hand-to-hand combat.

Hair was also important in ancient Rome, and many of the practices followed those of the Greeks. Hair and beards were curled with oven-heated curling irons or tongs. Wealthy Romans sprinkled gold dust on their hair, and some dyed their hair red with a caustic soap. The most popular hairstyles for men were short and brushed forward with arranged curls. Beardless faces remained in vogue. Women's hairstyles most often were parted in the center, waved or curled over the ears, and left hanging in long curls or put up in chignons or buns.

Hairstyles and hair care technology reached new levels in the Middle Ages. In addition to cutting hair, barbers also performed as surgeons, a practice previously attended to by monks, priests, and other clergymen. From the late twelfth to the mid–eighteenth centuries, barbers were called "doctors of the short robe" and practiced tooth pulling, bloodletting, and treatment of abscesses as well as hair care. Technological advances made during this period focused chiefly on methods of tinting and dyeing hair. Beards and longer hair for men, often cut in a pageboy, as well as braids for women returned in fashion. Both men and women wore a variety of hats or ornate and unusually shaped headdresses to cover the head. Blond hair continued to be admired, and many colored their hair using bleaching formulas that included a variety of ingredients such as henna, horse flowers, saffron, eggs, and calf kidneys. Hair and scalp conditioner was made from lizards boiled in olive oil and egg whites to give hair body and stiffness. Curling papers, crimpers, and ribbons were tools used to produce desired styles.

The Renaissance, particularly in England, was characterized by a variety of beards, mustaches, and hairstyles for men made popular by Henry VIII. The popularity of Queen Elizabeth I's red hair inspired her subjects to imitate her by wearing red wigs or by dyeing their hair red and shaving their hairlines to give the appearance of a high forehead. The high collars and starched ruffs of this period led to elaborate hairstyles that were sometimes constructed over a wire frame to achieve a heart shape around the head. Titian, the Italian artist, popularized a reddish-blond hair color in his paintings, and women who wanted to imitate the color applied mixtures of alum, sulfur, soda, and rhubarb to their hair and sat in the sun to let it dry. In Renaissance France, flowers were pulverized into a powder, then applied to the hair using glue.

In general, men's hair got longer and beards shorter in the mid-1600s. Wigs (especially for men) gained in popularity again, getting progressively fuller, curlier, and more exaggerated. The upswept feminine style of the Elizabethan period became flatter on top, with soft curls worn over the ears and forehead and the back of the hair pulled into a bun and decorated with ribbons, pearls, nets, feathers, and flowers. Among the less affluent, hairstyles were considerably more conservative. Among the Puritans, hair was worn short and uncurled, typically under a cap. In the New World, regulations were passed in 1634 forbidding all students at Harvard to engage in the fashion extremes of hairdressing, including long hair, lovelocks, and hair powdering.

Men wore wigs throughout Europe and the American colonies during the sixteenth century. Fashion shifted from bulky, exaggerated, and full-bottomed wigs to wigs of a smaller proportion—most frequently worn with a ponytail tied with a ribbon. Women's hairstyles became quite elaborate by the middle of the sixteenth century. False hair and padding were used to build hairstyles to great heights, and extravagant decorations—from jewels, flowers, and feather adornments to models of ships, bird cages, or entire flower gardens—were worked into them. The messy practice of dressing and powdering the hair took several hours, and wealthy people set aside a special room (powder room) in their homes for this purpose. White powder made of flour, starch, or plaster of Paris was used for most hair and wigs, but pastel-colored powders were also favored. Poorer people who desired to maintain fashionable hairstyles wore plain wigs and powdered them with sawdust. The wigs worn today by British judges, barristers, and royal coachmen are based on eighteenth-century styles.

The French Revolution brought an end to extravagances in hairstyle with a return to the classical Greek styles. Hairpins, clips, and tortoiseshell combs became popular hair ornaments. Wigs were rarely worn in the nineteenth century, and men once again wore facial hair in a wide range of styles. Treatments and cures for baldness were concocted of substances as varied as bear grease, beef marrow, onion juice, butter, flower water, sulfur, or mercury. The most widely used hair preparations were Macassar oil (imported from Ujung Pandang on the island of Sulawesi) and brilliantine (a derivative of coconut oil), whose function was to give hair shine. The simplicity of the smooth, center-parted styles worn by women in the Victorian era lasted until the 1870s, when the Parisian hairdresser Marcel Grateau created a new, deep, natural-looking wave by turning a curling iron upside down. The Marcel wave remained popular for almost half a century and helped usher in a new era of women's waved and curled hairpieces that were mixed with the natural hair. In addition to Grateau's curling iron, William Gillette invented the safety razor in 1895. Barbers now concentrated on cutting hair and trimming beards and mustaches, and a new age of at-home grooming practices began.

The invention of electricity changed hairdressing forever. In 1906, London hairdresser Charles Nestlé invented the permanent-wave machine that used heat generated by electric current passing through coils of wire onto which the hair was rolled. It took up to ten hours to complete the process of waving the hair and united fashion with twentieth-century technology. In 1907, Parisian chemistry student Eugène Schueller founded the company L'Oréal by developing a dye to cover gray hair with natural-looking colors.

The end of World War I saw changes in hairstyles for both men and women. A short haircut for women, called the bob, gained popularity because of its practicality for women working outside the home. Meanwhile, the rise of the film industry made a profound impact on hairstyles. Women around the globe quickly adopted the styles and colors of Hollywood actresses. Jean Harlow and Mae West were the first to popularize platinum blonde waves, and Shirley Temple redefined what it meant to have curly hair. The 1940s were more encouraging for brunettes like Katharine Hepburn and Hedy Lamarr, and red hair became popular once again in the 1950s with Lucille Ball.

For men, fashions did not change quite as radically or as rapidly as in the first half of the twentieth century, and short hair prevailed from the military influence of the two world wars. Elvis Presley and James Dean helped change that with long sideburns and the shiny pompadour. But it was the Beatles who radically changed men's hair fashion. Suddenly, long hair was in. The 1960s saw changes in women's hair as well. The return of straight hair and the asymmetrical haircuts created by English hairdresser Vidal Sassoon were a radical departure. Since the 1970s, there has been a wider acceptance of variety in hairstyles for both men and women—from straight and free-swinging to naturally curly to spiked-up "punk" hairdos.[72]

Corsets

Volumes have been written about the role of fashion in the enhancement of physical appearance for both men and women. A review of the history of clothing is beyond the scope of this chapter, but there is one article of clothing that appears to have been so blatantly used to alter physical appearance that it warrants some discussion.

The widespread use of corsets started in England and France in the sixteenth century with the arrival of Catherine de Medici in the French court as bride of Henry II (1533). Corsets would wax and wane in popularity in Europe over the next three hundred or so years, but it was clear that they had a dramatic effect on the shape of a woman's body. Much more than simply trimming, the idea was to compress the torso into an hourglass shape. The net effect was a dramatic accentuation of the bust and the hips, and this change was not lost on writers and poets of the time, nor were the deleterious health consequences of wearing them overlooked by physicians.[73] Numerous books and articles were written warning women of the dangers of corsets including, but not limited to, increased probability of stillbirths and abortion, and high risk of contracting consumption.[74]

In the eighteenth century, wood and metal stays were replaced by lighter whalebone as the favored construction material.[75] No matter the material, however, corsets restricted mobility and placed considerable stress on the rib cage and internal organs. A particularly tightly laced corset could exert pressure of eighty pounds per square inch under the right conditions. Clearly, such pressure would have been sufficiently high to do considerable damage, given that the recommended inflation pressure for a professional soccer ball is nine to twelve pounds per square inch, and for the tires on a car about thirty.

The erect formal posture associated with wearing a corset was identified with moral integrity and social prosperity.[76] Interestingly, the corset was the source of two widely used expressions. "Loose" women were those who loosened the stays in their corsets or removed them completely, while those characterized as "straitlaced" did just the opposite. The popular press eagerly printed accounts of the havoc wrought by corsets. In Paris in 1859, there was a report of the death of a young, particularly attractive, twenty-three-year-old woman. An autopsy revealed that she had died because her liver had been punctured by three of her ribs, which had been displaced gradually over a number of years by wearing a corset.[77] Consumption was the fashionable disease of the mid–nineteenth century and was considered to be due to corset wearing. Sufferers were believed to be in a continuous state of sexual excitement, and their compulsion to copulate supposedly led to their deaths. Now this is an article of clothing that should be treated with real respect.

PERMANENT BODY ALTERATION AND THE QUEST FOR BEAUTY

Skull Deformation

Hairstyles and makeup are temporary and in most cases have a short useful life span. However, in many cultures there exist a wide variety of practices that are directed toward permanent alteration of physical appearance. Cranial deformation is common and includes alterations of the skull, lips, teeth, tongue, nose, eyes, or ears. Cephalic deformation is the best-documented form, largely because skeletal remains in a number of prehistoric sites clearly show its presence. Flattening of the skull was produced by constant pressure of small boards or other flattened surfaces tied against an infant's head. Cranial deformation is known from all continents except for Australia. It is rare in sub-Saharan Africa and apparently absent from South India and Oceania.

Obviously, cranial deformation must be accomplished when an individual is young, when the skull bones are still pliable and the sutures have not closed. This potential for deformation is a direct result of our evolutionary history. Natural selection favored individuals with larger and larger brains, while the biomechanical constraints of bipedalism made expansion of the birth canal impossible beyond a point. Natural selection favored individuals whose offspring possessed deformable crania to meet the demands of childbirth. So we can thank our evolutionary history for the physiological and mechanical plasticity that allows for the deformation of skulls.

Cranial deformation has been used in Africa, Peru, Greenland, France, and by native North Americans. What is significant is not so much the widespread nature of the practice but the plasticity of the human skull. Profound changes in the shape of the human skull can be induced if started early enough in development, with little serious damage to the brain. Only in its most extreme expression are there reported cases of epilepsy associated with cranial deformation.[78]

Piercing and Scarification

In addition to reshaping skulls, there are a number of other practices that involve scarring, cutting, piercing, filing, or removing specific body parts in the service of beauty or prestige. Among the Botocudo of Brazil, young women inserted larger and larger disks into a hole in their lower lip until they had produced a "spoon bill." Among the Sara of the Ubangui-Chari region of central Africa, the labrets (lip disks) were worn in pairs by the women.[79] These labrets were so large that they actually interfered with speech. At the most extreme they were reported to be twenty-nine inches in diameter.[80] The famous explorer David Livingstone noted that

> the women seen at a distance and in profile seemed to be holding two saucers between their teeth. Eating was very difficult, and for all practical purposes those with the labrets were unable to speak. The labret was grossly uncomfortable; yet it was a mark of honor and people regarded it so highly that they hesitated at nothing to get it and to celebrate its attainment lavishly.[81]

Some have suggested that the use of lip disks was started by Sara men in order to make the women so unattractive that the slave traders would not be interested in capturing them.

In addition to wearing such lip plugs, people have pierced their ears, punctured their nasal septum, pulled teeth, and filed teeth all in the service of beauty. One of the most widespread beauty practices is scarification of the face and body. Scarification is particularly prevalent among dark-skinned groups, presumably because the process produces highly visible, often pinkish marks against the dark skin. The process involves lifting and cutting the skin to produce wounds that are as deep as the recipient can stand, followed by the application of an irritant preparation that inhibits healing, and finally rubbing black paint or wood ash into the wounds to produce a pronounced and highly visible scar. Over time, the same wound is opened and the process repeated to produce raised, hypertrophied scar tissue. Among the Tiv of central Nigeria, women undergo a series of scar-ification rituals to mark their maturation. Scars are made at puberty, at the onset of menstruation, and after weaning of the first child.[82]

All of this may seem to be quite removed from what goes on in Western society, but deliberate scarring is not unknown in the West. Instead of thinking simply about the aesthetic properties of scarification, consider that scars may be a mark of age, prestige, bravery, or status, and hence attractive under some circumstances. In Germany, for instance, duels of honor were sanctioned by the military code of justice until World War I and were le-galized again in 1936 under the Nazis. The Fascist regime in Italy also en-couraged dueling. The Mensur (student duel) is still a feature of German university life as a form of sporting event. Most German universities have long-established Verbindungen (fighting corps) with strict rules, secret meetings, distinctive uniforms, and great prestige.

In duels held under the auspices of the Verbindungen, a method of swordplay was employed that was distinct from that of normal fencing, dur-ing which participants obtained scars on the head and cheek that were prized as marks of courage.[83] The *renommier schmiss*, or bragging scar, was a mark of social status, and it indicated that you had been to a university where dueling societies were critical to social life. These duelists sub-scribed to an ethos of exaggerated masculinity in which skilled swordplay and fleetness of foot were less important than bravery. Using saber-type weapons, the antagonists—wearing body padding and iron goggles—faced each other for dozens of brief rounds, with each man defending against five whacking strokes and then taking five cuts of his own. There was nothing that could remotely be called fencing in these encounters; they were tests of fearlessness and endurance. The goals of the duel were to show courage, stand your ground, and not inflict a wound on your opponent but be wounded yourself. The point was not to injure your opponent but to show that you could take your whacks.

Women responded positively to these proudly displayed scars, which the combatants often soaked with beer or stuffed with horsehair to increase their size and prominence. The scars showed you had courage and an ed-ucation, characteristics that made you an attractive mate. Even if it was an ugly, pronounced scar, women were attracted by its symbolic nature. In or-der to preserve at least some of their unblemished looks, young men usu-ally accumulated the scars on the left side of their faces so that from the right profile they looked uninjured. Rather than being impediments, such scars were attractive decorations that actually enhanced an individual's prestige and status.[84]

Tattoos

If your skin is not dark and you would rather not duel, don't despair. People with lighter-colored skin have not been left out in the race to devise some permanent form of body decoration. Rather than scarification, tattooing seems to be a technique of choice for decorating the face and the body of light-skinned peoples. Tattooing has a long and distinguished history among modern humans.

Darwin, in *Sexual Selection and the Descent of Man*, commented that "not one great country can be named, from the Polar regions of the North to New Zealand in the South, in which the aborigines do not tattoo themselves."[85] Tattooing has been found among the ancient Thracians, Assyrians, Gauls, and Britons. Tattoos are depicted in carved figures from Europe over eight thousand years old, in ancient Egyptian paintings and figurines, as well as in Japanese pottery over three thousand years old. By around 2000 B.C., the practice of tattooing spread from the Mideast to Asia and the Pacific and was well established as a permanent form of body decoration by 1000 B.C.[86]

Among the Maori of New Zealand, tattooing was developed into an art form and served as a marker of high status. Their tattoos were inscribed with a serrated bone or shell adz dipped in pigment (the design was literally chiseled one-eighth inch into the skin with an adz that was struck with a mallet). The process was extremely painful and was associated with elaborate rituals that dealt with changes in status.[87]

As an art form, tattooing is practiced with consummate skill in Japan and has been for hundreds of years. In Japan, the practice of tattooing has been in and out of favor. In the mid–nineteenth century, tattooing was outlawed by Emperor Meiji, but for the preceding two hundred years it had been popular among all social classes. About this time, Japan began to come into regular contact with the West, and one of the consequences of extensive Western contact was that the underground practice *hori* (meaning to engrave) flourished among European and American sailors, merchants, and visiting dignitaries.[88] By the early twentieth century, tattooing had become an undesirable art form, particularly for individuals of the upper class. Laborers, artisans, criminals, entertainers, and especially firefighters were the most frequent clients of Japanese tattooists. Today it is still true that Japanese tattooing is the most ornate, intricate, and colorful in the world.

Tattooing in Western society can be traced to the original inhabitants of the British Isles. In fact, Briton is derived from a Breton word meaning "painted various colors." Caesar is reported to have said that the Britons were colored blue[89] and carried designs that made them "frightful to look upon in battle."[90] This kind of face painting, not really tattooing in its truest sense, was readily adopted by the Roman legions and flourished in the military until it was banned in the third century by Constantine, who said it was a violation of God's handiwork.

Centuries later, during the Crusades, soldiers had themselves tattooed with Christian religious icons to ensure a Christian burial should they die in a foreign land. Until the rise of Christianity, tattoos were commonplace in Europe but were banned by the Christian church. Tattoos had virtually disappeared when they were "rediscovered" by European sailors who came into contact with American Indians and Pacific Islanders. Some explorers returned home wearing tattoos; they also brought back drawings of decorated islanders and Indians—and occasionally the islanders and Indians themselves, who were exhibited at fairs and circuses.[91] The newly popular tattoos were favored mostly by working-class Europeans, but they enjoyed a brief spurt of popularity among upper-class men and women in England in the late nineteenth century.

John Smith visited North America in 1593 and reported that the native inhabitants "have their legs, hands, breasts and faces cunningly embroidered with diverse marks, such as beasts and serpents artificially wrought into their flesh with black spots."[92] Modern tattooing in the West began in the late eighteenth century with the voyage of Captain James Cook to the South Pacific. Not only did Cook and his men return with numerous tattoos attained during their journey; Cook also introduced the Tahitian word *ta-tu*, meaning to strike or mark, in preference to the original term "pricking." Officers and sailors of the H.M.S. *Endeavor* received tattoos to commemorate their voyage. Over the next two centuries or so tattooing achieved some degree of popularity among working-class men. Sailors continued to strongly favor tattoos as marks of completion of their various voyages, passing the equator, or sailing around the world, to mention just a few.

By the mid–nineteenth century, a number of Europeans were making a living by exhibiting themselves and their elaborate tattoos to the public as well as to gatherings of prestigious medical associations. One of the first to be completely tattooed was a Frenchman, Joseph Cabri, who was exhibited publicly and presented to several crowned heads in Europe. In spite of his notoriety, he died a pauper in Valenciennes in northern France in 1818. Another famous decorated entertainer was Englishman John Rutherford, who achieved some notoriety not only for his whole-body tattoos but also for his tales of his capture and four-year stay among the Maoris of New Zealand (1816–1820). During his incarceration, he was lavishly decorated with the tribal markings of a Maori chief.

One of the most flamboyant exhibitionists was an Albanian known as Constantine, Captain Constantine, or Prince Constantine. He claimed to have been forcibly tattooed by "Chinese Tartars" in Burma. His body was covered with 388 small, highly detailed, etched figures. As evidence of the popularity and interest these men stirred in the public, Constantine was managed by the famous circus entrepreneur P. T. Barnum. The first tattooed lady, La Belle Irène, was tattooed over not only the most visible parts of her body; even the soles of her feet and the spaces between her toes were covered. These exhibitions, as well as the folklore associated with these world travelers, started a fad in Europe by the late nineteenth century.[93]

In the early part of the nineteenth century, sailors, craftsmen, the military, as well as members of the aristocracy were involved with tattooing. It was during their travels to Asia and the Pacific that members of the aristocracy were tattooed as a mark of their exotic experiences.[94] The tattooing craze rapidly spread to the United States and was readily accepted by the upper class, while others expressed strong objections. There are injunctions in the Old Testament against tattooing: "You shall not make any cuttings in your flesh on account of the dead or tattoo any marks upon you."[95] These admonitions were recognized by a large portion of the Christian community.

Martin Hildebrand was the first professional tattooist in the United States, and during the Civil War he is reported to have tattooed thousands of the Confederate and Union troops. Shortly after the war, Hildebrand opened a shop in New York where, along with a few others, he began to ply his trade. Tattooing fell out of favor among the American upper class in the early part of the twentieth century. Stories of the spread of venereal diseases from the lack of sanitary conditions in tattoo establishments fueled the decline of tattooing in the United States, and by the 1920s tattooing was regarded as a deviant practice.[96] Still, the Depression forced many people to become heavily tattooed in order to earn a living by being displayed as curiosities in circuses and sideshows.

By the mid–twentieth century, tattooing was firmly established as a deviant practice in American society. Tattooists were seen as largely unskilled, dispensing their decorative products in unsanitary conditions from dingy shops in urban slums to marginal, socially deviant, and dangerously unconventional members of society. Since the mid 1960s, tattooing has enjoyed a revival; rather than being seen as some sort of deviant activity, tattooing is now seen much more as an art form, accepted by all social classes. An estimated twenty million Americans were believed to sport tattoos in 1996, with the number almost double that today.

For some people a tattoo promised invincibility in war. For others tattoos protected against sickness or misfortune, or offered safe passage into heaven or the afterworld. And for still other people, they furnished a visible badge of rank or membership in a certain group: witness the number of firefighters from all over the United States working at Ground Zero who have been tattooed with images commemorating the disaster of 11 September. Tattoos have served as a way of advertising one's emotional (or philosophical) attachments. Most commonly, however, tattoos have been and still are used for decoration.

From an evolutionary perspective, it is unlikely that we will find a specific gene that codes for tattooing behavior, so we must ask what the benefits of any type of permanent body decoration are to the individual. In parts of Polynesia and Micronesia, tattoos were used to denote a change in maturational and social status. Tattoos were used in India to mark people of difference castes. Tattoos have been used not only to separate people into classes or categories; they have also been used to show wealth and status, and hunting or fishing prowess.

Today tattoos are often used to show group affiliation. Gangs in the United States are known to use tattoos not only as a mark of identification but also as a rite of initiation. Interestingly, tattoos have also been used to mark a lack of achievement or even disgrace. Tattoos have been used to mark slaves very much as cattle were branded in the American west.[97] Jews were tattooed when they were incarcerated by the Nazis during World War II.

Tattooing was also done in order to make individuals more attractive to the opposite sex. In many cultures, tattooing is a mark of beauty. Tattoos were not restricted to one sex. In various cultures only males were tattooed, while in others only females, and in some both. Among the Ainu, a group that lives on Hokkaido, the northernmost island in the Japanese chain, tattooing was used to produce an extremely exaggerated mouth on young girls. Tattoos have also had high symbolic value, showing the virility of a male, for example, or providing protection from an enemy.[98]

The pursuit of beauty through such a permanent means of body decoration is not without costs to the bearers. In addition to the costs of tattoos in terms of time, money, pain, and suffering, there are particular health risks associated with the sterility and hygiene of the tattooing procedure. As early as 1862, the dangers of tattooing were recognized when M. Ernest Berchon published a report that called attention to the individuals who had contracted syphilis while being tattooed. He also noted a number of cases of infection necessitating the amputation of limbs as a result of nonhygienic tattooing procedures. In 1877, James Kelly, a tattooist working in Philadelphia, was tried and convicted of infecting scores of men with syphilis.

There are approximately twenty-two different diseases that can be transmitted by needle-stick injuries including syphilis, malaria, and tuberculosis. The disease most frequently transmitted by needle contamination is hepatitis B. In the United States, the Centers for Disease Control and Prevention estimates that twelve thousand health care workers become infected with hepatitis B each year from needle sticks. About 25 percent of those develop acute hepatitis, and about 10 percent will go on to become carriers. Hepatitis B is highly infectious, in fact, about forty times more so than AIDS.[99] A number of cases of tattoo-related hepatitis B infection have been reported but no HIV cases. The most dramatic occurrence was in New York City in 1960, when thirty cases of hepatitis B and one fatality were blamed on tattoo artists using unsterilized needles and contaminated pigment. In addition to hepatitis B, a number of other skin conditions can be developed as a consequence of tattooing, but none of these are fatal. Today, tattooing is illegal in New York City, Connecticut, Florida, Massachusetts, Oklahoma, South Carolina, Vermont, and Albuquerque, New Mexico.

Another beauty technique with potentially lethal side effects is practiced by the Padaung women of upland Myanmar, who are famous for their stretched necks. Necks are stretched as much as fifteen inches, with approximately four thoracic vertebrae pulled up into the neck by coiled brass neck rings. Girls of this group begin wearing brass neck rings at an early age. Initially, five brass rings are placed around the neck with additional rings added gradually, until each individual has twenty-four rings around her neck.[100] In addition to the neck rings, brass rings are put on the arms and around the ankles until a woman is carrying an additional fifty to sixty pounds of brass, all the while engaging in her normal daily activities. The effect of the neck rings is to increase the length of the neck by stretching all the musculature in the neck as well as the soft tissue surrounding the vertebral column, while driving the clavicle down so far that it overlaps with the first few ribs. The consequences of this lengthening of the neck are so dramatic that if the brass rings were removed, the women would not be able to hold their heads up.[101]

Cosmetic Surgery

It is clear that body decoration and ornamentation are a panhuman characteristic, and one of the critical aspects of our appearance is the possibility that it can be altered. The earliest records of cosmetic surgery come from India over two thousand years ago. The Hindu surgeon Sushruta described reconstructing a patient's nose from the skin on the cheek. During the Middle Ages, anatomists like Vesalius, Fallopius, and Paré wrote of the dangers of cosmetic surgery. The first printed work specifically on cosmetic surgery was *De Chirurgia Curtorum* (1597) by Gasparo Tagliacozzi of Bologna, Italy, who is often considered the father of modern plastic surgery. In his opus on surgery, he included twenty-two detailed woodcuts illustrating facial surgeries in progress. Tagliacozzi pioneered the Italian method of nasal reconstruction in which a flap of skin from the upper arm is transferred to the nose.

The greatest triumphs in plastic surgery come in the repair of disfigurements caused by traffic accidents and warfare as well as congenital conditions. Not only is the physical deformity repaired; these surgeries often give new psychological hope to people who have suffered profound psychological and physical disasters. Birthmarks, naevi ("port-wine stains"), hyperteorism (Grieg's disease),[102] and cleft palate all respond to surgery. In addition, a variety of less severe conditions can also be treated successfully. Humped noses, bat ears, as well as receding chins can be eliminated by plastic surgery.

Cosmetic surgery as an area of professional expertise in the United States is first associated with Charles Conrad Miller (1881–1950), who opened his practice in Chicago in 1903. Up until then, cosmetic surgery had been performed by quacks and charlatans. Miller not only was a capable and daring surgeon, he was also an excellent writer who, by the sheer power of his prose, convinced countless numbers of midwesterners to come to him to correct the "imperfections of life."

Miller wrote numerous articles that appeared in local, regional, and national medical journals. He was known for operations to correct crow's-feet, eyelid, forehead, and nasolabial wrinkles, thick and thin lips, large, small, and protruding ears, as well as the introduction of an original method for creating dimples. Miller's work was not universally accepted by his colleagues, and, in fact, some questioned his use of paraffin injections to correct what he called a "hump nose" as well as in the treatment of an inguinal hernia. His description of the use of sponge rubber and gutta-percha,[103] which he ground up in an ordinary "spinach grinder" as a substitute for paraffin, did not instill confidence in other members of the medical profession.[104]

Regardless of Miller's questionable reputation, cosmetic surgery was off and running. Now I want to make a critical distinction here, and that is between plastic surgery to repair disfigurements that are the result of catastrophic events or genetic deformation and surgery performed simply to enhance the appearance of an otherwise normal-looking individual. Reconstructive surgery is performed on abnormal structures of the body caused by congenital defects, developmental abnormalities, trauma, infection, tumors, or disease. It is generally performed to improve function but may also be done to approximate a normal appearance. Cosmetic surgery, on the other hand, is performed to reshape normal structures of the body in order to improve and enhance the patient's appearance and self-esteem.

Cosmetic surgery has enjoyed a colorful history in the United States and is intimately linked to changing ideas of beauty. In the early twentieth century, "natural" standards of beauty vied with artificial ones. Both early feminists and progressive thinkers advocated natural beauty and argued that living right, eating right, and thinking right would lead to the development of character. On the other hand, advancing technology in the cosmetic industry, coupled with widespread publicity and marketing strategies, driven by the consumption-for-pleasure ideology, tended to undermine the Victorian strictures against the artificial enhancement of beauty and spelled the demise of the natural ideal.

This firmly held belief in the perfectibility of the human form is attributable in no small part to the journalists of the day. The *New York Daily Mirror* held a contest in December 1924 to find the homeliest girl in New York. First prize in the "Homely Girl Contest" was a free surgical makeover and an opera audition. Rosa Travers, a sweatshop worker, was the winner and was given a makeover by Dr. W. A. Pratt, owner of the Pratt Feature and Specialty Company of Brooklyn. In addition to the contest, the paper featured a first-person account about how looks had affected her life.

The broader idea of self-improvement has deep roots in American society. In the seventeenth and eighteenth centuries, self-improvement was couched largely in religious terms,[105] but in the nineteenth century, as a result of the increasing secularization of society, moral self-improvement became redefined as better character. In the early part of the twentieth century, this emphasis on character improvement gave rise to broader concerns about the development of personality. Personality embraced a wide array of beliefs, attitudes, and behaviors but, more than that, was the presentation of self. While character was internal, personality was subject to all sorts of external trappings, including but not limited to appropriate speech, dress, and manners. The division between moral character and personality began to become quite fuzzy.

It was frequently said that personality gets jobs, wins friends, draws members of the opposite sex like a magnet, and keeps spouses in love, but, most important, personality could be acquired. The often heard admonition was that you must be willing to work for what you want, and that you needed willpower, courage, and determination to achieve your goals. Once diets, exercise, cosmetics, and hairstyles were accepted as weapons in the battle to improve physical attributes, the distinction between physical appearance and personality became blurred. It was not difficult then for people to accept cosmetic surgery as a legitimate form of self-improvement. Nowhere is this more clearly seen than in advertising.

Beginning in the 1920s, advertisers began the relentless promotion of the "first impression." Americans were told that first impressions could spell success or failure. The theme of a good first impression was exploited not only in presentation of self but also in one's house, one's car, and even in one's toilet paper. Even the chaos caused by the Great Depression did not dim American enthusiasm for perfection, especially when it came to appearance. The *New Republic* estimated that in 1931 Americans spent over $750 million on cosmetics and that beauty shops employed more than 100,000 women.

The Great Depression had an interesting effect on the way cosmetic surgery was viewed. The Depression had a devastating economic impact on all Americans, setting the stage for an increasingly competitive view of life. It is not surprising that plastic surgeons of the day began emphasizing the competitive advantage that appearance could give in the economic struggle. Surgeons enthusiastically championed the benefits of enhanced appearance, but their admonitions would have had little effect were it not for the realization by many Americans that life was tougher in the 1930s than it had been in the decade before.

The final piece of the puzzle linking morality and, ultimately, plastic surgery fell into place in the 1920s. Plastic surgeons touted the power of changes in appearance in overcoming psychological disabilities. Changes in appearance were linked to profound changes in personality and personal happiness. As a discipline, psychology was becoming increasingly more visible at this time. Terms like "repression," "unconscious," and "Oedipus complex" began to appear regularly in newspapers and conversation. Freudian psychoanalysis was a popular topic of discussion, but it was not Freud who scored a direct hit on the psychology of cosmetic surgery. It was Viennese psychologist Alfred Adler who formulated a new idea, the inferiority complex, as an explanation of much of human behavior. Unlike Freud's work, which was deep, dark, and forbidding, the inferiority-complex idea held much appeal, particularly in the face of the Depression. Psychologists all over the country observed that, prior to the Depression, the mood of the country was optimistic, but the Depression left major scars on the American psyche.

By the 1930s, the notion of the inferiority complex was used to explain all sorts of maladies, ranging from unruly behavior in children to failed marriages, suicide, and bankruptcy. The link between physical appearance, mental health, and social and economic success had been firmly forged. That link had been indelibly etched on the American psyche. Now cosmetic surgery had the underlying psychological basis that it had previously lacked.

One of the most widely publicized cosmetic surgical procedures in the early twentieth century was performed on Fanny Brice, the famous vaudeville actress, comedienne, and star of Flo Ziegfeld's Follies. She had a nose job, which was performed in a New York hotel room by Henry Junius Schireson, a surgeon with a dubious educational background and a less than illustrious career. At the time of the surgery, Brice proclaimed her appreciation for the surgeon's work. Brice's nose job attracted more press attention "than any other medical event until the birth of the Dionne quintuplets."[106]

The first public face-lift was conducted at the 1931 International Beauty Shop Owners Convention, although little is known of the details. *Popular Science Monthly*, in November 1937, featured the photo spread "New Noses in 40 Minutes." The photographic essay showed twelve photographs taken at different intervals throughout the entire surgical procedure. The transformation of an unshapely nose could be done on one's lunch break. Popular magazines touted the virtues of plastic surgery, such as its lifting the psychological burden of some deformity of appearance carried by the sufferer.

The treatment of injuries to soldiers after World War II convincingly demonstrated the psychotherapeutic importance of plastic surgery. Many victims of war wounds and burns regained use of the injured parts of their bodies and a return to most of their prewar appearance. After World War II, some historians have argued, there was an integration of the ideas of beauty with those of youth.

The discovery of lanolin and the invention of cheap aerosol dispensers changed the cosmetic industry forever. The American quest for youth[107] and beauty was fueled by the explosion of the cosmetic industry, and by 1956 American women spent $2.5 billion on cosmetics, beauty products, and weight reduction aids. By the mid 1960s, cosmetic surgery was openly and thoroughly discussed in the popular press. Face-lifts were characterized as better for your morale than a trip to Europe and actually a better value in the long run.[108] Efforts were made to characterize cosmetic surgery patients as serious, hardworking women, not simply narcissists or hedonists.

The 1960s also witnessed the bridging of the gender gap for cosmetic surgery. Early in the decade, stories began to appear in a number of popular magazines (*Coronet, Forbes, Gentleman's Quarterly*) about the economic benefits to men who had face-lifts. The economic justification for cosmetic surgery for men continued to increase through the years, and by the 1970s, 15 to 20 percent of the patients getting face-lifts were men. Vidal Sassoon, Elvis Presley, and Senator William Proxmire would not be classed together on almost any other criteria one might choose, but all had eyelid surgery. Another odd couple, Milton Berle and Walter Mondale, both had rhinoplasty.

The fascination that the public has with cosmetic surgery continues today with popular daytime talk show host Oprah Winfrey's "Plastic Surgeons Create Their Perfect Wives" spectacle in 1992. Lip augmentations have been shown on live television, and cosmetic surgery procedures performed on a European performance artist were videotaped and later broadcast on commercial television, as was one of Joan Rivers's face-lifts. Medical reporters and television news anchors across the United States had plastic surgery and documented it for all to see, as well as had their employers pay for the procedures.

Many popular personalities have undergone cosmetic surgery, but none is any more recognizable than Michael Jackson. His pale, almost transparent skin that is stretched over high cheekbones, his tiny, chiseled, pointy nose, his deeply cleft chin, and his deeply lined eyes almost defy gender and ethnic classification.

American culture is characterized as having a tendency to change rather than cope, and to alter rather than endure. The discomfort often expressed over Michael Jackson's surgeries is based partially on their extent and cost. Elizabeth Haiken, author of *Venus Envy: A History of Cosmetic Surgery*, observed that "Jackson looks less like his presurgery self and more like Elizabeth Taylor than either patients or surgeons in the previous generations would have believed possible."[109] Noted pediatrician Benjamin Spock commented, "He seems partly child, partly adult, partly masculine, partly feminine; he seems to be a person for all ages and sexes. I don't see him doing any harm, but I am sure he's not doing any good either."[110]

While the details are unclear and stories are conflicting, Jackson admits to two rhinoplasties and an artificial cleft in his chin. Others say that Jackson has had four rhinoplasties and two "touch-ups." Still others suggest that he must also have had cheekbone implants, permanent eyeliner, and surgical reduction of the lower lip. To many, Jackson's facial overhaul is troubling because of a larger general feeling in society of declining access to high-quality health care and the stark contrast his appearance makes to what could ever be expected in the average American. This is among the most extreme and blatant examples of conspicuous consumption.

CULTURE OF YOUTH AND COMPETITION

Today in the United States, elective cosmetic surgery is routinely performed to "improve" the appearance of individuals. The United States is an image-conscious culture, and, unlike earlier times, the choice of radically altering one's appearance is now possible and only modestly expensive. Americans live in a culture that is pathologically concerned with youth and beauty. It is not a culture that values all members equally. Those that more closely conform to the current abstract notion of what is attractive are more successful. Americans not only value beauty and reward it—they have also raised the recognition of beauty to the level of a competitive encounter.

One of the first public organized contests that attempted to define beauty was the Miss America competition. Margaret Gorman was crowned the first Miss America in 1921, and for the first time beauty was explicitly made a major criterion in the competition. Thus, the era of organized mate competition as we now know it had begun.

Today, Miss America, Ms. America, Miss Teenage America, Miss U.S.A., America's Most Pampered Princess (infant), Miss Sweet Pea, All American Woman Plus,[111] Miss Palmetto State, Miss Plump Universe, Miss Magnolia State, Miss Grits Festival, Ms. Maxi America, Miss National Alternative Lesbian, and Little Miss Lake Sam Rayburn, to mention just a few, are all contests where participants are pitted against one another, not in some test of strength or prowess, but in a contest over appearance. Not only does this competitive beauty culture often involve individuals in the prime of their sexual attractiveness, but it also embraces individuals who range in age from newborn infants to senior citizens. Today, it is not difficult to find a pageant just for your age and marital status.[112] One of the consequences of this concern with beauty is that beauty gets equated with good and ugliness with evil. It is no accident that, in much of the children's literature, beauty and good are synonymous. For example, in *The Wizard of Oz*, Glinda the

Table 3-1 Summary of Cosmetic Surgery Procedures

Procedure	Number 1990	Number 1999	Number 2000	% Change[a]	Number Males	% Males	Cost ($)	Age	2000 Expenditure ($)
Abdominoplasty	6,864	87,802	93,756	6.78	4,688	5.00	5,310	43	497,844,360
Blepharoplasty	85,520	423,719	465,177	9.78	88,384	19.00	3,130	50	1,456,004,010
Botox	n/a[b]	623,588	730,787	17.19	65,771	9.00	434	42	317,161,558
Breast augmentation	41,918	255,254	288,044	12.85	0	0.00	4,556	31	1,312,328,464
Breast lift	4,560	54,568	62,100	13.80	0	0.00	4,658	39	289,261,800
Breast reduction	6,323	101,228	114,497	13.11	8,015	7.00	5,430	38	621,718,710
Chemical peel	n/a	1,432,071	1,897,508	32.50	284,626	15.00	1,121	45	2,127,106,468
Face-lift	36,981	224,216	244,370	8.99	31,768	13.00	5,416	54	1,323,507,920
Fat injections	n/a	297,635	262,008	-11.97	31,441	12.00	1,573	44	412,138,584
Forehead lift	26,436	132,783	145,881	9.86	18,965	13.00	3,353	50	489,138,993
Gynecomastia	5,654	30,030	27,941	-6.96	27,941	100.00	3,060	33	85,499,460
Hair transplant	57,714	216,351	222,497	2.84	177,998	80.00	3,661	43	814,561,517
Laser resurfacing	n/a	310,188	300,572	-3.10	42,080	14.00	3,117	50	936,882,924
Liposuction	71,632	599,430	672,793	12.24	114,375	17.00	3,816	40	2,567,378,088
Malar augmentation	9,570	18,793	21,481	14.30	5,370	25.00	2,513	37	53,981,753
Microdermabrasion	n/a	433,502	551,398	27.20	43,996	7.98	100	39	55,139,800
Otoplasty	6,136	25,595	26,881	5.02	11,290	42.00	2,699	25	72,551,819
Rhinoplasty	89,615	171,442	155,052	-9.56	44,965	29.00	4,109	33	637,108,668
Sclerotherapy	n/a	442,209	445,211	0.68	17,808	4.00	311	41	138,460,621
Totals	448,923	5,880,404	6,727,954	8.61	1,019,481	15.15		41	14,207,775,517

Source: American Academy of Cosmetic Surgery, 2001.
[a]Percent change refers to changes from 1999 to 2000.
[b]Data are not available.

Good Witch of the North is attractive, while the Wicked Witch of the West, with her green complexion, is less than stunning.[113]

So, returning to the subject at hand, cosmetic surgery is defined as a surgical procedure on an otherwise normal face or body part in order to enhance the appearance and benefit the patient emotionally and psychologically. In other words, it is possible to change one's appearance through modern biomedical technology so that it more closely conforms to some ideal standard defined by culture. Increasingly these standards of beauty do not make any allowance for the normal aging process and treat aging as a degenerative condition and a deformity that can be corrected.

Rhinoplasty (nose job), rhytidectomy (face-lift), blepharoplasty (eyelid adjustment), otoplasty (ear flattening), gynecomastia surgery (male breast reduction), malar augmentation (cheek implants), microdermabrasion (removal of epidermis), liposuction (removal of localized fatty deposits), and blepharopigmentation (eyeliner tattoo) are all becoming commonplace procedures.

Table 3-1 shows the distribution of cosmetic surgeries performed in the United States in 2000. Almost 6.8 million cosmetic surgery procedures were performed that year, an overall increase of slightly less than 9 percent from 1999. Elective cosmetic surgery can no longer be thought of as just for women, for over a million procedures were performed on men in 2000. Cosmetic surgery is big business, with Americans spending over $14 billion in 2000. By any standards, this is a growth industry.[114]

The most popular procedure, by a substantial margin, was the chemical peel. As shown in table 3-1, while there are no data for 1990, the technique has been around for a while. It became an approved surgical technique in the early 1960s. Prior to this, chemicals had been applied to the skin in an attempt to rejuvenate it, but this was not done in a medical setting. The chemical peel process of today uses a variety of procedures. The least complicated is often advertised as the "lunchtime fix." Glycolic acid or other alphahydroxyl acids (AHA) are applied to the face of the patient, and, over the days following application, the outer layers of skin slough off. Five to seven applications are recommended to obtain the best results.

Another somewhat stronger chemical, trichloroacetic acid (TCA), is also used for removal of unwanted skin. Like glycolic acid, a TCA peel takes about fifteen minutes to apply, and more than one application may be required to obtained the desired results. Finally, for those requiring removal of a significant number of layers of skin, the strongest chemical peel—phenol—is applied. Unlike the other procedures, a phenol peel requires anesthesia and results in permanent lightening of the skin. All the procedures require patients to exercise caution when they are exposed to sunlight. Complications from chemical peels are, generally, disturbances in skin pigmentation, and pigmented facial blemishes are often darker after the procedure. Persistent facial redness sometimes occurs after a chemical peel but usually resolves within three weeks.

The second most popular procedure in 2000 was the intramuscular injection of botulinum toxin type A (Botox®). Botox is a neurotoxin that is produced by the bacterium *Clostridium botulinum*, which, if encountered in large quantities, will cause a type of paralysis called botulism. Botox is a neuromuscular-junction blocking agent, and it acts by blocking the release of the neurotransmitter acetylcholine. It does not permanently damage the nerves of the muscles; it just disrupts the messages between them. When small amounts of Botox are injected into specific muscles, the muscles become paralyzed. The most common use is to reduce frown lines on the forehead. Injection of Botox into the corrugator muscle between your eyebrows makes it impossible for you to frown. Once one of the corrugator muscles is paralyzed, the frown line gradually fades away. Unfortunately, a Botox treatment is not permanent, and within three to five months repeated treatment is necessary. Botox is also used for treatment of other wrinkles, including crow's-feet. To date there have been no major complications from the use of Botox, but it has only been in widespread use since the late 1980s.

Liposuction followed closely behind Botox in the number of procedures performed in 2000. Liposuction has changed over the last two decades since its introduction in France in the late 1970s. Early on, liposuction involved making incisions in the skin, scraping the fat cells, and suctioning them away. More specialized tools and techniques have been developed over the years that have dramatically increased the success of the procedure and reduced discomfort significantly for the patient.

Now there are basically two liposuction techniques that are used. Standard suction-assisted lipectomy (SAL) is a procedure in which the area to be suctioned is injected with large volumes of a dilute solution of a local anesthetic, sterile saltwater, and a mild stimulant. When the tissue is firm to the touch (tumescent), the procedure can begin. By flooding the area to be suctioned prior to the procedure, larger volumes of fat can be removed with lower risk of fluid and blood loss and concomitant electrolyte shifts.

The most recent liposuction procedure uses ultrasonic energy to assist in the removal of subcutaneous fat. Instead of simply using suction to remove the subcutaneous fatty deposits, a rapidly vibrating probe is inserted into the tissue and ultrasound is applied. The result is that the fat cells are ruptured, and they release their fatty acids into the intracellular space. The rupture of the fat cells, or adipocytes,[115] "emulsifies" the fat, creating a milky yellow fluid that can be removed by suction.[116]

As surgical procedures go, cosmetic surgery is considered by many to be quite safe. Risks range in severity from the relatively benign (rippled and dimpled skin surface) to the serious (abdominal perforation, skin sloughing, and significant blood loss). In addition, there is the obvious danger from anesthesia.[117]

According to a report by the California Medical Board, between 60 and 100 deaths in California alone were due to complications from the 149,000 liposuction procedures done in 1997. More recently, concern was expressed in an article in the *Anesthesia Patient Safety Newsletter* about the rate of complications from liposuction procedures conducted in doctors' offices, considering that 25 percent of all procedures requiring anesthesia are not performed in hospitals. A review of mortality from liposuction procedures performed by board-certified members of the American Society of Plastic and Reconstructive Surgeons published in *Plastic and Reconstructive Surgery* from 1994 to mid-1998 showed a mortality rate of 19.1 per 100,000 population. An earlier survey (1997) by the same society showed an almost identical rate (20.3). Given that mortality rates from HIV are 5.2 per 100,000, colon cancer 20.8, breast cancer 15.4, motor vehicle accidents 15.2, and homicide 5.9, a mortality rate of 20 per 100,000 is astonishing. Remember, these data are from licensed plastic surgeons and do not take into account non–board-certified plastic surgeons, dermatologists, primary care physicians, emergency physicians, anesthesiologists, ophthalmologists, dentists, and cosmetologists, all known to have performed liposuction procedures.[118] In a recent paper in the *New England Journal of Medicine*,[119] clinicians reviewed deaths during liposuction procedures and concluded that considerable attention should be paid to the amount of lidocaine and fluid used in the common tumescent liposuction procedures because of the potentially lethal side effects.

In addition, there are other potential risks linked to cosmetic surgical procedures, such as impaired vision and blindness due to eyelid lifts, eyelids that do not close completely because too much skin was removed from the lower lid, and permanent skin discoloration resulting from laser treatment of wrinkles.[120] Given these data, it seems that the rather benign view that cosmetic surgery is relatively risk-free is unfounded. It has been said that, compared to private cosmetic surgery clinics, veterinary clinics and funeral homes have more specific standards.[121]

Liposuction, like all elective cosmetic procedures, is not covered by insurance. Since patients who elect this type of surgery must pay for it out of their own pockets, this becomes a desirable procedure from the physician's perspective because of the significantly reduced administrative costs. As most physicians will tell you, treating patients is not the problem; keeping up with all the bureaucracy associated with health care is what bogs the system down. Cosmetic surgical procedures can be done in a private office and do not generally require hospitalization. Consequently, physicians are largely free from intrusion by regulating bodies and associated bureaucracy. Since procedures are typically done in the physician's office, engaging in an active marketing campaign is possible, especially for less expensive noninvasive procedures such as chemical peels and Botox injections. Taken together, these factors make cosmetic surgery very lucrative, so much so that a wide variety of medical specialists have taken up the calling. In fact, in some states, dentists and podiatrists are practicing limited cosmetic surgery.

While women are the primary targets for these kinds of surgeries, men are also getting cosmetic surgeries in increasing numbers. In 2000, the most popular procedure for men was the chemical peel, followed closely by hair transplants, then liposuction. Interestingly, men accounted for over 15 percent of all procedures in 2000. The choices of procedures are consistent with the societal emphasis on youth that equates baldness, wrinkles, and obesity with advanced age, weakness, and unattractiveness. Men spent over $2 billion on cosmetic surgeries in 2000, and the trend is not expected to change in the next few years. It is important to remember that in no cases were these surgeries clinically required.

One of the interesting aspects to the question of whether cosmetic surgery is beneficial is the extent to which plastic surgeons are actually helping patients. Or are they instead furthering a system that drives women and men to attempt to achieve almost unattainable standards of beauty? Does a person dissatisfied with his or her appearance have a right to correct an aesthetic defect? Is this a right that should be supported by private insurance as well as the government? Should it be available only for those who can pay? Is the allocation of scarce resources for training plastic surgeons in cosmetic surgery something that should be continued when there is a scarcity of physicians willing to treat the underserved portions of our population? Might it be possible to slow the juggernaut of fashion by imposing luxury taxes on the fashion, cosmetic, and related industries?[122] Perhaps it is time to look at the extent to which federal support of the cosmetic and fashion industries fuels the quest for youth and beauty. Is such support appropriate?

Now from the point of view of an evolutionary anthropologist, it is easy to understand why people would exploit the services of plastic surgeons if they can afford to do so. The whole enterprise is driven by the deeply held psychological predisposition to attract mates. While most would not put it quite so boldly, it is obvious that if there is a way to make oneself more attractive, particularly if costs are low, then many people will engage in the practice. The power of this psychological predisposition to increase one's attractiveness should not be underestimated.

People are not only willing to spend extraordinary amounts of money on these procedures, but they may also be committing themselves to future procedures, since most are not truly permanent. In addition, people usually do not seriously consider the physical dangers that are associated with these procedures, even though the data speak volumes to the contrary.

Today, American and much of Western society is caught in an extraordinary spiral. On one hand, we have the evolved physiology to accumulate fat either on our buttocks and thighs or around our midline, as a consequence of repeated exposure to food shortages in our evolutionary past. We have the evolved predisposition to engage in risky behavior via cosmetic surgery to enhance our attractiveness. Couple with that a culture that venerates youth, and you have a recipe for trouble. Now stir in, if you will, a large helping of biotechnology and continuous exposure to cultural models of beauty that are beyond the capability of most mortals, and you have a fascinating collision between biology and culture. As I have tried to convince you, most of us are engaged in all sorts of futile pursuits in search of the fountain of youth and are locked in mortal combat with the battle of the bulge. But this is a part of what it means to be human.

Is there a way out? Well, first of all, psychotherapy for the masses might help to enhance our self-acceptance and emphasize the idea that we do not have complete control over our bodies or what we look like. That is obviously impractical, but still we must come to recognize that we are products of a long evolutionary history and have strongly ingrained behaviors, tastes, and preferences that stood our ancestors in good stead for thousands of generations.

Second, we could begin to redefine what we consider attractive. Rather than deify youth, we could all bleach our hair gray and apply wrinkle-inducing crèmes. We could enlist the aid of cosmetic surgeons to enhance the signs of aging with such procedures as fat injections around the waist, buttocks, and thighs. Rather than forehead and breast lifts, we could have breast and forehead drops, eyelids could be made to droop, and we could surgically inscribe wrinkles on our faces. Of course, such efforts would likely have an uphill battle in completely redefining what our culture defines as beautiful.

CONCLUSION

Our culture and biology are on a collision course when it comes to attracting a mate. We have highly evolved psychological predispositions to engage in a variety of mate-attracting behaviors, and these predispositions were evolutionarily successful in the Paleolithic. The specific pattern of these behaviors is orchestrated by our local culture and, as I have tried to show, can be fitness-reducing rather than fitness-enhancing. The most dramatic examples of this conflict occur when the costs of the behavior are high. While there are good evolutionary reasons why advertising something about one's desirability as a mate should be expensive in order to discourage cheaters, modern technology has allowed wholesale modification of the human appearance into something that often has little biological reality.

The quest for beauty has a long and checkered history in the course of human experience. Individuals have died in pursuit of its elusive qualities, while others have spent an entire lifetime in its pursuit, never achieving their goal. It is our fundamental biology that has set the stage for people to go to such extremes and endure such high costs in the pursuit of beauty. In the vocabulary of evolutionary biology, finding a mate and producing successful offspring, who themselves have offspring, is the only way of winning the evolutionary game. Our culture and all the technological advancements that we enjoy have not diminished the intensity of the competition, nor are they likely to. One can only hope that, if we understand a bit more about our evolutionary history, culture may not push us as blindly in this quest for unattainable, unrealistic, everlasting youth and beauty.

4 Fat, Diet, and Evolution

All the things I really like to do are either immoral, illegal, or fattening.
—Alexander Woollcott

THE NATURE OF OBESITY IN WESTERN SOCIETY

Obesity is a medical condition characterized by the storage of excess body fat. The human body naturally stores fat tissue under the skin and around organs and joints. Fat is critical for good health because it is a source of energy. When the body lacks the energy necessary to sustain life processes, it can mobilize fat. Fat also provides insulation and protection for internal organs. Therefore, from a health standpoint, fat is good. However, the accumulation of too much fat in the body is associated with a variety of health problems. Studies show that individuals who are 20 percent or more overweight run a greater risk of developing diabetes, hypertension, heart disorders, stroke, arthritis, and some forms of cancer.

There is an epidemic of obesity in the United States. "The proportion of the population that is obese is incredible. If this were tuberculosis, it would be called an epidemic."[1] Obesity (defined as being over 30 percent above ideal body weight) in the population increased from 12 percent in 1991 to 17.9 percent in 1998. The highest increase occurred among individuals aged eighteen to twenty-nine, people with some college education, and people of Hispanic ethnicity. By region, the largest increases were seen in the South, with a 67 percent increase in the number of obese people. Georgia had the largest increase—101 percent.[2]

According to surveys conducted in 1977–1978 and 1994–1996, daily caloric intake increased from 2,239 kcal to 2,455 kcal (kilocalories) in men and from 1,534 kcal to 1,646 kcal in women. Eating frequency is influenced by a number of environmental changes, not the least of which is the growth of the fast-food industry and the accessibility of its products, the proliferation of a greater variety of foods with higher caloric content, and the increase in the types and marketing of snack foods. Coupled with the increase in eating frequency is the decrease in overall calorie-burning activity: children watch more television; many schools have done away with physical education classes; neighborhoods lack sidewalks for safe walking; household chores are assisted by labor-saving devices; and automobiles are used for all except the nearest travel destinations.[3]

However, in the United States, there is a long history of dieting and concern about fat.[4] The first American weight watchers were disciples of Sylvester Graham in the early nineteenth century. Born in West Suffield, Connecticut, in 1794, Graham was the youngest child in a family of seventeen. Graham led a troubled life and was considered by many to be either an eccentric and a genius or a crackpot. At age thirty-six, he abandoned the ministry to devote full time to becoming a professional reformer. First, he campaigned against alcohol; then he began a crusade against the ill effects of a poor diet. For Graham, the concern was not about fatness or obesity; rather it was about being resilient and robust, not gluttonous.[5] The moral imperative of living the correct lifestyle had been laid down. Graham's moral imperative served to galvanize the opinion of many against those who were obese, or those who did not lead the prescribed virtuous lifestyle.

The mid–nineteenth century saw the decline of society's acceptance of fat. Corpulence was on the way out, and, by the end of the Civil War, there was real disrespect for fat. The turn of the twentieth century saw the rise of slimming techniques (e.g., fasting, calorie counting) that dealt with fatness as a threat to the system, a moral as well as a physical danger.[6]

But the weight-watching culture really came alive in the Roaring Twenties. American society became preoccupied with the healthy body, and one of the major components of physical health was weight. People began an obsession with weight, and that led to the realization on the part of scale manufacturers that money was to be made by charging the American public to weigh themselves. Platform scales appeared on street corners, in drugstores and grocery stores, and in public areas all over the country. For a penny, one could step onto the scale and get weighed.[7] The penny scale became the index of health. The culture of thinness had arrived in the United States.

Health problems related to obesity have long been recognized, but only recently have empirical data been collected that allow a more complete understanding of the confounding effects of multiple factors that contribute to obesity (e.g., genetic, cultural, socioeconomic).[8] The relationship between obesity and hypertension is well documented. Individuals who are overweight are approximately three times as likely to be hypertensive. Meanwhile, the "good" (high-density lipoprotein, HDL) cholesterol that is implicated in risk reduction for coronary heart disease and resistance to certain types of serious infections is lower in the obese. And the "bad" (total and low-density lipoprotein, LDL) cholesterol is elevated. High levels of LDL can result from a genetic deficiency in LDL receptors or from a modern high-cholesterol diet that stimulates body cells to reduce their LDL receptors, once their cholesterol needs have been met from a diet high in fat, or a diet that is particularly rich in foods that are high in saturated fats.[9]

Obesity is also implicated in diabetes, coronary heart disease, gallbladder disease, respiratory disease, and some forms of cancer. In addition to the obvious physical costs of obesity, there are psychological costs as well. Obese individuals are often seen as lacking willpower, self-esteem, and intelligence. Obesity and sexuality are intimately connected according to some,[10] with obesity negating any feelings of self-worth and attractiveness. Most troubling, however, are studies that demonstrate that health professionals have a negative view of obesity.[11] In a study of medical students' attitudes about obese people, the students as well as practicing clinicians characterized the obese as weak-willed, ugly, awkward, unpleasant, worthless, or bad.[12] Obesity was rated as the fifth most negative characteristic of patients in another study.[13] Socioeconomic status is also implicated in obesity. In a study of over ten thousand women aged sixteen to twenty-four, researchers found that overweight women were less likely to marry than other women, had lower incomes, had incomes that were more often below the poverty level, and had completed fewer years of school than nonobese women.[14]

Obesity places a substantial financial burden on American society. In a report published in the journal *Obesity Research*, investigators calculated the various component costs of obesity. They found that in 1995, the year for which they had the most adequate data, the total cost of obesity was $99.2 billion, with slightly over half of that going for direct medical costs. Almost 6 percent of total health expenditures is attributable to obesity.[15]

The National Health and Nutrition Endpoint Survey III,[16] conducted in 1996, found that approximately 35 percent of women and 31 percent of men age twenty and over were obese, as compared with about 25 percent of the total population in 1980. About one-quarter of children and adolescents were obese.[17] In the United States overall, one-third of the adult population was obese, compared with one-quarter in 1980. Obesity is more prevalent in minorities,[18] particularly women of minority groups—nearly half of African American and Mexican American women are obese.[19] While clinicians, researchers, and epidemiologists all agree that obesity is wide-spread, part of the discrepancy in the actual statistics presented arises from differences in the operational definitions of obesity. But for our purposes, the question of concern is not how to "cure" obesity but rather how to understand it as an outcome of culture and biology working in conflict.

USDA GUIDELINES, WHAT WE EAT VERSUS WHAT WE ARE EVOLVED TO EAT

Concern over nutrition and the quality of foods has been around for centuries. Governmental regulations relating to food can be traced back to the first English food law, the Assize of Bread, enacted in 1202 by King John (1199–1216), which prohibited the adulteration of bread with ground peas or beans. The Adulteration of Food and Drugs Act passed in 1872 in England was one the earliest pieces of legislation that formally dealt with consumer issues. Other early legislation is recorded in the United States, where there have been a number of laws addressing consumer health and safety:

- The Adulteration of Food and Drugs Acts of 1848, 1890, and 1906 set the stage for governmental intervention and control of consumer products.
- The Gould Amendment of 1913 was the first law dealing with labeling the contents of packages of food.
- The first Food Standards were issued in 1929.
- The Food Additives Amendment was enacted in 1958, requiring manufacturers of new food additives to demonstrate their safety.
- The Food and Drug Administration was established in 1988.
- The Nutrition Labeling and Education Act of 1990 required that all packaged foods bear nutrition labeling.[20]

Dietary recommendations also have a long and rich history in the United States. The United States Department of Agriculture (USDA) issued the first dietary guidelines in 1916.[21] In 1941, the Committee on Food and Nutrition of the National Research Council published a pamphlet, "Recommended Dietary Allowances (RDA)." This was the first attempt to specify optimum levels of protein, calcium, iron, vitamins A, B, C, and D, riboflavin, and nicotinic acid. Other dietary recommendations followed:

- The Basic 7 Food Guide (setting guidelines for milk, vegetables, fruits, eggs, meat, cheese, fish and poultry, and butter) was popular from 1943 to 1955.

- The Basic Four Guide (milk, vegetables and fruit, bread and cereal, and the meat groups) was in vogue from 1956 to 1979.
- The Hassle-Free Food Guide, a remake of the Basic Four Guide, was distributed from 1979 to 1984.
- The Daily Food Choice Guide and the Food Guide Pyramid were introduced in 1984, and versions of both are in use today (see fig. 4-1).

The Food Guide Pyramid offers recommendations on the numbers of servings that individuals should consume every day of six basic food groups. Figures 4-2 and 4-3 show a comparison of USDA recommendations, the actual consumption of these food groups in the United States,[22] and consumption during the Neolithic (essentially the diet that we were evolved to eat).[23] While reconstructing the exact diet of our ancestors is difficult, scientists have suggested that our ancestral diet contained no refined sugar, no processed grains, little meat, and substantially more fruits and vegetables than consumed today. Our reconstructed Neolithic diet is at odds with the USDA dietary recommendations in several critical areas:

- number of servings of milk, yogurt, and cheese (USDA 2.5,[24] Neolithic 5.0);

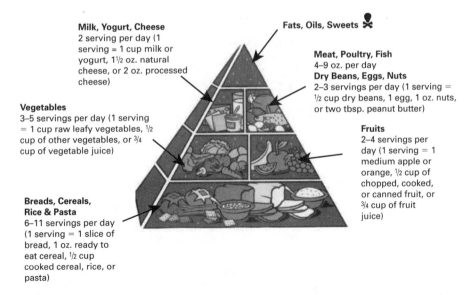

Milk, Yogurt, Cheese
2 serving per day (1 serving = 1 cup milk or yogurt, 1½ oz. natural cheese, or 2 oz. processed cheese)

Fats, Oils, Sweets

Meat, Poultry, Fish
4–9 oz. per day
Dry Beans, Eggs, Nuts
2–3 servings per day (1 serving = ½ cup dry beans, 1 egg, 1 oz. nuts, or two tbsp. peanut butter)

Vegetables
3–5 servings per day (1 serving = 1 cup raw leafy vegetables, ½ cup of other vegetables, or ¾ cup of vegetable juice)

Fruits
2–4 servings per day (1 serving = 1 medium apple or orange, ½ cup of chopped, cooked, or canned fruit, or ¾ cup of fruit juice)

Breads, Cereals, Rice & Pasta
6–11 servings per day (1 serving = 1 slice of bread, 1 oz. ready to eat cereal, ½ cup cooked cereal, rice, or pasta)

Fig. 4-1. USDA recommendations for good nutrition. The Food Pyramid shows the different foods groups identified by the USDA and the recommended number of servings per day.

- number of servings of bread, cereal, rice, and pasta (USDA 8.5, Neolithic 7.0) and fruit (USDA 3.0, Neolithic 5.75);

whereas meat consumption and sugar consumption are remarkably close for the two estimates.

Critics will question the calculation of the Neolithic diet, and rightfully so. Estimates of food consumption by our evolutionary ancestors are notoriously difficult at best, but these data are the best currently available.

Perhaps the most interesting question is how the USDA derived its recommendations and how these recommendations have changed over time. What we have today is an attempt to offer a recipe for good nutrition that one would assume was based on solid scientific data. However, the economic significance of these guidelines should not be overlooked, for they serve as the basis for *all* federally funded food programs in the United States.

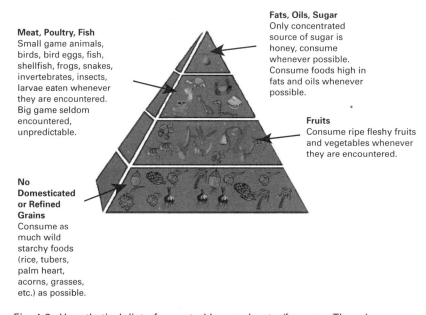

Meat, Poultry, Fish
Small game animals, birds, bird eggs, fish, shellfish, frogs, snakes, invertebrates, insects, larvae eaten whenever they are encountered. Big game seldom encountered, unpredictable.

Fats, Oils, Sugar
Only concentrated source of sugar is honey, consume whenever possible. Consume foods high in fats and oils whenever possible.

Fruits
Consume ripe fleshy fruits and vegetables whenever they are encountered.

No Domesticated or Refined Grains
Consume as much wild starchy foods (rice, tubers, palm heart, acorns, grasses, etc.) as possible.

Fig. 4-2. Hypothetical diet of ancestral human hunter/foragers. There is disagreement among specialists about what might have been our ancestral diet (see text for further details), but the main point is that wild foods were exploited in ways that are unexpected based on today's diet. Wild starchy and leafy foods were a major portion of the diet, as were fruits, nuts, and berries. Small game was a chief source of protein, along with insects, larvae, and invertebrates. Honey was the only concentrated sugar source encountered. © E. O. Smith, MMI.

Given the potential economic as well as health importance of the USDA guidelines, it is not unexpected that there has been some controversy over conflicts of interest between members of the Dietary Guidelines Advisory Committee (DGAC), the panel of experts that formulates the recommendations, and various food industries. Recently, the USDA forfeited its right to appeal a recent court decision that it (USDA) violated federal law for hiding conflicts of interests of the DGAC. The Physicians Committee for Responsible Medicine (PCRM), a Washington, D.C.–based research and advocacy group, brought the suit against the USDA. "Having advisors tied to the meat and dairy industries is as inappropriate as letting tobacco companies decide our standards for air quality," noted the PCRM president, Neal D. Barnard.[25]

Does this mean that we should ignore dietary guidelines developed by the USDA? Of course not, but these difficulties highlight some of the

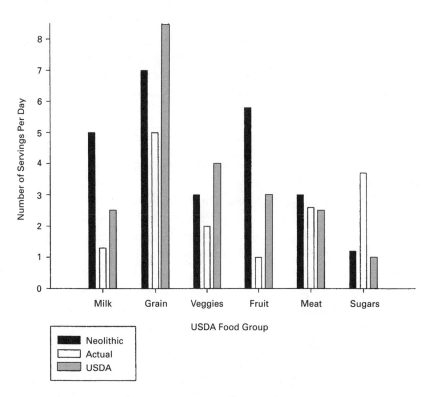

Fig. 4-3. Graphical representation of the USDA recommendations, what we actually consume, and our hypothetical evolutionarily stable (Neolithic) diet. © E. O Smith, MMI.

problems associated with broad-based policy guidelines for nutrition. There are certain segments of the population that should not follow the guidelines too closely. Those who are lactose intolerant, those who have a family history of cardiovascular disease, and people who are diabetic are particularly vulnerable to problems that might arise from following the USDA guidelines. An evolutionary perspective simply suggests that we should not rely so heavily on cereal grains, reduce the amount of processed sugar we consume, and eat much more fruit.

To many, the USDA guidelines seem like a commonsense approach to nutrition, and something that is well within our capacity to achieve. But let us not be too quick to judge the dietary practices of others without first considering the difficulty many people have in obtaining such a wide variety of foods. It is tempting to think that the full range of foods suggested by the guidelines is nothing more than a quick trip away from our dinner tables. Not so for a substantial proportion of the population. Many people in the United States who have poor nutrition have low incomes as well, and they have to rely on public transportation to get to food stores. When they arrive, they find that purchasing foods that follow the guidelines is not cheap. Moreover, for many of the recommended foods, preparation time is significant. For many people, it is easier and cheaper to rely on fast food and snacks that are notoriously low in the nutrition hierarchy.

THE FAST-FOOD CONNECTION

Fast food[26] is a non sequitur to some, or, depending on your view, it is an oxymoron, a dietary staple, or even a pox on the land. Like it or not, the fast-food industry has over the last half of the twentieth century made a profound impact on many dimensions of Western society. What started out as a couple of hamburger stands in southern California has insinuated itself throughout society. Generations of Americans have grown up on Big Macs, Happy Meals, and Whoppers. Fast food is now served at most airports, zoos, high schools, elementary schools, colleges, and universities, on cruise ships, on airplanes, and, believe it or not, in hospitals.[27] Fast food is big business. In 1970, Americans spent about $6 billion on fast food, while in 2000 the tab was more than $112 billion.[28] To put that in perspective, the National Defense Budget for 2000 was $270 billion.[29]

Of the fast-food corporations, McDonald's is the largest. One out of eight workers in the United States has worked at McDonald's at some time in his or her life. McDonald's hires over a million people per year, more than any other public or private organization in the United States.

McDonald's is the largest owner of retail property in the world, the largest purchaser of beef, pork, and potatoes, and is second only to Kentucky Fried Chicken for chicken.

McDonald's is a cultural icon immediately recognizable by people anywhere in the world. A survey of American schoolchildren found that 96 percent could identify Ronald McDonald. The only fictional character with a higher recognition rating was Santa Claus.[30] In fact, over forty-three million people are served every day in over twenty-eight thousand restaurants in 120 countries. This means that McDonald's serves the equivalent of approximately 15 percent of the population of the United States, or almost 1 percent of the world's population, every day. Such staggering statistics have not gone unnoticed in the corporate world. Children are the focus of a significant part of McDonald's advertising budget.[31] McDonald's advertising is so successful that 87 percent of all children in the United States between the ages of six and nine come into its stores once each month.

On 11 April 1997, the McDonald's Corporation embarked on one of the most astonishing marketing campaigns ever staged, when it announced that one of ten different Teenie Beanie Baby dolls was to be given away with the purchase of a Happy Meal. In the following ten days, McDonald's sold nearly one hundred million Happy Meals and completely exhausted the projected thirty-five-day supply of Teenie Beanie Babies. Happy Meals are marketed to children from three to nine years of age, so that means that there were approximately four Teenie Beanie Baby Happy Meals sold for every American child in the targeted market.[32]

In addition to the approximately three hamburgers and four orders of french fries the average American consumes each week, the average twelve- to nineteen-year-old male consumes fourteen cans of soda pop per week. Soda consumption is only slightly less for females.[33] More than 85 percent of all fourteen-year-old males report drinking at least one soft drink every day, and one-third consume at least three twelve-ounce soft drinks each day.[34] The classic Big Mac, fries, and a Coke constitute over 60 percent of the total caloric intake of a two thousand kcal per day diet, with more than 80 percent of the recommended daily allowance of fat and slightly less than 60 percent of our daily sodium requirement.

To what can we attribute the unbridled success of a business that sells foods that are largely of questionable nutritional value? There are complicated cultural reasons that have to do with globalization, the expansion of democracy and capitalism, massive sociopolitical changes, and economic reform, but I argue that one of the main reasons that McDonald's food sells so well is that it tastes good.

Now, what is it about the food that makes it taste good? It tastes good to us because the staples of the menu are foods particularly rich in substances that, in evolutionary terms, have high survival value. Fats, sugar, and salt are dietary components that are essential to life. Each performs a different physiological function, but all are necessary. In our evolutionary past, each of these compounds was in short supply, and, when our ancestors were able to obtain them, those who were able to consume the largest quantities or process the compounds most efficiently were favored by natural selection. Our highly developed taste preferences for salt, fat, and sugar are the consequence of this strong selective pressure. No wonder an astonishing number of customers flock to McDonald's and other fast-food restaurants every day. Couple that with the amount of other prepared foods with similar nutritional content, and it is easy to see that our diet is not guided by rational healthy choice but, in large part, by what tastes good.

How Do They Make It Taste So Good?

The bench standard in the fast-food industry is the french fry, and McDonald's took this imported dish and not only perfected the taste but also mechanized its production. Here is the fundamental problem: the plant source for french fries is an erect herb (*Solanum tuberosum*), a member of the nightshade family, known more commonly as the potato.[35] Producing french fries might appear to some to be a simple process, but there are many factors that manufacturers have to take into account to produce a french fry that is crispy on the outside, not too dark, and of the perfect consistency on the inside.

One of the first challenges that "fry meisters" must overcome is a biochemical problem. Consumers are finicky about their french fries, and they do not want them to be too dark. Potatoes are rich in sugars, so if you take a fresh potato, slice it, and fry it, it will turn dark on the outside long before the inside is cooked. Sugar is the culprit. Actually, it is the reaction of reducing sugars (glucose and fructose) with amino acids in the presence of heat that forms the darkened color.[36] Consumers do not want soggy french fries either. Potatoes are mostly water, and water is the cause of soggy french fries. The process of creating a french fry involves removing as much of the water as possible and replacing the water with fat. When a potato is fried, the heat of the oil turns the water inside the potato into steam, which causes the hard granules of starch to swell and soften. At the same time, the outward migration of steam limits the amount of oil that invades the interior of the potato, preventing the fry from becoming too greasy. Oil is concentrated on the surface, where it turns the outer layer of the potato brown and crisp.[37]

One of the most astonishing technological innovations was the "potato computer," a device developed for McDonald's by former electrical engineer Louis Martino. One of the problems that had to be overcome in the production was how precisely to determine the optimal cooking time for a batch of fries.[38] In addition, there was concern about curing time, potato variety, water content, and techniques of the frying process itself, as well as the type of oil used in frying. Ray Kroc, the founder of McDonald's, used animal fat to cook his fries, specifically a mixture of approximately 7 percent soy oil and 93 percent beef tallow. This mixture gave the fries their unique flavor.[39] Kroc insisted that all his franchised restaurants use his specially formulated beef tallow oil, which he called Formula 47.[40] Previously, making high-quality french fries had been an art. The potato computer, the hydrometer, and the curing bins made it a science.

McDonald's and other fast-food restaurants used these techniques for preparing french fries until 1990, when widespread concern grew over the deleterious effects on cholesterol levels of cooking with saturated fat.[41] McDonald's responded to the public concern by changing from Formula 47 (saturated fat) to vegetable oil. In reality, this was little improvement because the vegetable oil was hydrogenated, which produces trans fatty acids. Trans fatty acids (unsaturated fats that have been hydrogenated) have adverse health effects that equal or exceed those associated with saturated fats. Saturated fats are those found in animal tissue and are implicated in a variety of chronic cardiovascular diseases. Trans fatty acids, however, have been shown to impede the body's ability to regulate cholesterol. So, while french fries may taste good, they are actually very efficient delivery vehicles for trans fatty acids.

Concern over the nonspecified content of the cooking oil used by McDonald's recently sparked a class action lawsuit by Hindu and non-Hindu vegetarians. In the face of impending litigation in both Washington and Texas, McDonald's admitted that its fries were cooked with vegetable oil that contained other "natural flavors." The FDA does not require companies to disclose the contents of special flavorings added to food, because that information is considered proprietary. The "natural flavoring" added to McDonald's is a beef derivative, a food item that offends vegetarians and violates the Hindu religion. While McDonald's now specifies the contents of its cooking oil on its company website, the lawsuit is in the courts. The plaintiffs contend that McDonald's has been guilty of deceptive advertising since 1990, when it started using "vegetable" oil for french fries. The announcement that it is not entirely vegetable prompted protests outside the corporate offices in New Delhi. In Bombay, protesters broke into a McDonld's and destroyed furniture and ceiling fixtures; later protesters covered Ronald McDonald with cow dung at another Bombay store.[42]

Why do we find the taste of food cooked with saturated fat or trans fatty acids so appealing? Well, the answer is a bit complicated. Humans characterize substances as tasting sweet, salty, bitter, or sour. We determine taste by signals sent from receptors in our mouths, called taste buds. Taste buds are not randomly distributed around our mouth, tongue, or pharynx but are located primarily on the tongue. The mucus membrane of the surface and sides of the tongue, the roof of the mouth, and the entrance to the pharynx are sites for numerous tiny projections called papillae. Each of the papillae has two hundred to three hundred taste buds. The papillae located at the back of the tongue are arranged to form a rearward-facing V and transmit the sensation of bitterness. Those papillae at the tip of the tongue transmit sweetness, while papillae on the sides of the tongue transmit saltiness and sourness. Therefore, substances that we put in our mouth stimulate taste receptors in different parts of the tongue, and those receptors send a message to the brain that interprets the electrical impulse as one of the four basic tastes. Simply put, this is how we taste substances. Now the next question is, why do we recognize only these four types of taste?

In our evolutionary past, the ability to recognize and differentiate the four basic tastes must have been highly favored by natural selection. The use of the senses is a basic theme in the evolution of vertebrates. Sight, hearing, touch, and smell provide animals with an immense amount of information about their environment. It is true that evolution has not elaborated the senses equally in all animals, and for some groups a particular sense has been extraordinarily elaborated or diminished. We humans are pretty good in terms of vision, but, when it comes to the chemical senses (smell and taste), we are not very impressive when compared to other animals or, for that matter, other mammals.

Salt

As we have seen, fat is crucial to survival, and there can be little doubt about the evolutionary history of our preference for it in our diet. In addition to fat, are there other substances that are so important to our basic physiology that we must have them? Salt is one of those essential minerals that we require for normal physiological function. The sodium and chloride found in salt, along with potassium, are electrolytes, substances that when in the proper solvent will conduct ions. Electrolytes are important in regulation of the body's fluid levels as well as the balance of acids and bases.

Considerable research has been conducted on salt by physiologists as well as psychologists. Physiologists are concerned with the regulation and management of electrolytes, while psychologists are interested in how we acquire our taste for salt. Is our preference for salt learned, and, if so, how is it accomplished? Alternatively, is our predilection for salt a consequence of our evolutionary history? Of course, the answer is both, but there are some other interesting answers to the recurring "why" question. Ecologists have noted that one of the consequences of heavy rainfall is the depletion of sources of sodium present in the food of land-dwelling herbivores. This is particularly true for animals that are far from the ocean. Female mammals may have developed a predisposition for salt due to the demands of pregnancy and lactation.[43] Mammals with a highly developed preference for salt would be favored in an environment where sources of salt were unpredictable. When salt was encountered, consumption of substantial quantities could have provided a reserve on which an individual might draw.

Once again, a comparative perspective is important to understand our craving for salt. A sodium hunger has been found in rhesus monkeys (*Macaca mulatta*) in laboratory studies.[44] It has been shown that, for humans in experimental situations, salt deprivation will cause foods to be less tasty and taste sensations less intense, and, when given salt-rich foods after deprivation, subjects reported the salty taste as pleasant.[45] A salty taste serves as a generalized marker for mineral deposits, since potassium-, calcium-, or zinc-deprived laboratory rats will ingest salt (sodium chloride).[46] In the natural setting, mineral-salt replenishment sites are called salt licks.

The point is, when animals moved into a terrestrial environment from an aquatic one, finding salt became a problem. In an aquatic environment, the problem is not too little salt but too much, so excretion becomes a problem. In a terrestrial environment, finding salt and storing it was the problem, since the body does not manufacture sodium.[47]

Archaeologists have found evidence of salt mining in deposits dated 8,500 years ago near Salzburg, Austria. Since the agricultural revolution, salt has been a part of human history and a valuable commodity. Wars have been fought over it, empires were founded on it and have collapsed without it, and people have sold their own children into slavery for a handful of it. Marco Polo wrote about the value of salt in his journeys through China in the eleventh century. In much of the Old World, salt was more valuable than gold. The use of salt as a measure of worth is clearly seen in the famous quote "not worth his salt" from Petronius in the *Satyricon*. Roman soldiers were paid in salt rations, called "salarium," or "salt money," which is the origin of the English word "salary." The Bible is replete with references to salt.[48] Salt was the most important state monopoly commodity until the industrial age and the production of fossil fuels.[49]

Common table salt is the most popular condiment in the world. And table salt (about 40 percent sodium) is the most common source of dietary sodium. The National Research Council of the National Academy of Sciences recommends a daily range of 1.1–3.3 grams of sodium, based on the amount of food consumed. For every one thousand kcal of food consumed, one should consume 1 gram of sodium, not to exceed the upper limit of 3.3 grams per day. Two-thirds of Americans consume 2.8–4.8 grams of sodium per day (i.e., 7 to 12 grams of salt). Salt intake exceeding 10 grams per day is considered excessive by clinicians.[50] That amounts to between eight and nine and a half pounds of salt per individual per year. Our closest salt-consuming competitor among European countries would be Italy, with about 10 grams per day.

In the 1980s, there was considerable concern over findings that salt was implicated in hypertension. One out of four African-Americans is affected by hypertension. The condition is one and one-half to two times more prevalent in African-Americans than in whites, and severe hypertension is five to seven times more common in African-Americans than in Caucasians. Similarly, disproportionate death rates are noted from cardiovascular disease or stroke. Blacks, older people of any race, diabetics, and obese individuals all show above-average sensitivity to dietary salt. For them, salt is a real health hazard. Nevertheless, all people in the modern Western world, not just a particular segment, eat more salt than their bodies need, so why is hypertension so unequally distributed among various ethnic groups? The often-heard answer is that certain ethnic groups must be more susceptible to hypertension because of some vaguely defined aspect of their genetic constitution. Current thinking suggests that the genetic explanation is not satisfying and is incomplete at best. Study of the dispersal of West African populations and their adaptation to local environments has demonstrated some interesting things about hypertension. First, it is well known that West African populations have among the lowest rates of hypertension in the world. It is also well known that immigrants from West Africa to the United States and Great Britain have among the highest rates of hypertension in the world; one study has included three populations of people of West African ancestry in three very different environments (rural Nigeria; Spanish Town, Jamaica; and Maywood, Illinois, a suburb of Chicago).

American and Jamaican blacks in this study share about 75 percent of their genetic heritage with Nigerians, so this was about as good a natural experiment as one could hope to conduct. While genes might account for some small proportion of the differences in hypertension, researchers found significant differences in the three groups. Of all the groups, Nigerians living in a rural village had the lowest incidence of hypertension (7 percent), while black Jamaicans were intermediate (26 percent) and American blacks were the highest (33 percent). These data suggest that factors other than genetics were playing a role in the expression of hypertension in these populations. Researchers concluded that, at best, genetic factors may account for 25 to 40 percent of the variation in blood pressure between people; the remainder surely must be due to a confluence of factors that relate to lifestyle, diet, exercise, education, and socioeconomic status. The point of this discussion is that, to explain ethnic differences in rates of hypertension, we have to look for a constellation of biological and also cultural factors that set the stage for the expression of clinical hypertension.[51]

Health concerns associated with high blood pressure, and the presumed causal link between hypertension and salt/sodium intake, have a long history. Chinese medicine has long postulated the deleterious effects of salt. Emperor Huang Ti (Di), the Yellow Emperor, in the twenty-seventh century B.C., wrote in his classic of internal medicine, *Nei Ching* (Jing), "If too much salt is in the food, the pulse hardens, the complexion changes, and tears make their appearance." Five thousand years later, however, the health effects of salt in the diet are still vigorously debated among clinicians. In the last two decades, a case has been made for the "salt hypothesis," which links high salt intake with high blood pressure.[52] As that hypothesis became widely accepted, salt was labeled a killer in the popular press, and broad public health initiatives were sought to restrict salt intake by the general population.

Recently, a growing body of data has caused some to question the harmful effects of salt in our diet. In 1995, the results of a four-year observational study in a population with high blood pressure suggested a surprising link between a *low*-salt diet and an increased risk of heart attack.[53] Advocates of the "salt hypothesis" found methodological problems with the study and tended to minimize its findings. An even greater furor was created by the publication in the *British Medical Journal* (*BMJ*) and the *Journal of the American Medical Association* (*JAMA*) of results of two studies concluding that little evidence existed for the correlation of salt intake and hypertension.[54] Almost unanimously, scientists and authorities called the *JAMA* study flawed, saying, among other things, that it was too limited, it included a subgroup of people who had normal blood pressure and were not fed consistent hospital-controlled diets, and it included trials that lasted too short a time.

As you can imagine, the jury is still out on the "salt hypothesis," so experts consequently have come to a middle-ground position. For individuals who are young and not hypertensive, modest amounts of salt are likely not very harmful, but for older individuals consumption exceeding the recommendations based on diet is unwise.

Fat

The human body requires carbohydrates, lipids, proteins, vitamins, minerals, and water for normal functioning. Most foods contain a combination of these nutrients. The remarkable part of the story is the amazing ability of our cells to covert one type of molecule into another; complex molecules can be broken down into smaller chemical units (catabolism), and larger chemical units constructed from smaller building blocks (anabolism). These activities are the basis of all of the body's chemical processes, collectively called metabolism.

The first part of the fat story concerns how the body utilizes lipids. In order to be of any use, lipids (which are not soluble in water) must be emulsified, suspended as small particles or droplets in a medium such as milk. Emulsification is particularly important in digesting concentrated lipids such as fish, liver, or animal fat. The stomach does a good job of mechanically emulsifying fats before they are passed into the intestine, where bile, a natural emulsifier, completes the process. The active ingredients in bile are complex salts that emulsify fats very much as soap does.[55] In contrast to other biological molecules, lipids are digested very slowly, second only to the most complex carbohydrates. Under normal circumstances, glucose, sucrose, and other simple sugars will be digested and transported to target tissues before any significant quantity of lipids enters the bloodstream. It may take six or more hours for adults to complete digestion and absorption of a meal that was high in fats.

In addition to lipids, it is important to understand the digestion of complex sugars. The liver is a truly amazing example of a biotechnologically sophisticated sugar-metabolizing machine. The sources for these essential nutrients are quite varied. Except for small amounts of glycogen (animal starch) in meats and lactose (milk sugar), all the carbohydrates we ingest are derived from plants. Sugars (both simple and complex) come from fruits, sugar cane, sugar beets, honey, and milk. Complex sugars are also found in grains, legumes, and root vegetables. From a strictly physiological standpoint, we require about one hundred grams of carbohydrates per day to maintain minimum blood glucose levels. When we consume only fifty grams of carbohydrates, tissue proteins (amino acids) and fats are used for energy.

The simple sugar (monosaccharide) glucose is produced by the liver from the digestion of carbohydrates, yielding two other simple sugars, fructose and galactose. Glucose is the major body fuel and is used to make adenosine triphosphate (ATP). ATP is not an energy storage cell. It is, however, the energy transport vehicle. Energy trapped in the tail of the phosphate cell is released when the action of a water molecule breaks the bonds between the phosphates (we call the process hydrolysis). It is true that many cells can and do use fats as sources of energy, but neurons and red blood cells rely on glucose. Glucose in excess of the amount needed for the production of ATP is converted into glycogen in skeletal muscle, the heart, and the liver as well as into fat in adipose cells.[56]

In addition to referring to lipids, a type of biological molecule, the word "fat" is also applied to living tissue. There are two kinds of fatty (adipose) tissue: white adipose tissue[57] and brown adipose tissue. The white fat serves as the primary repository for fat in the mammalian body and hence the primary energy reserve. On a pound-for-pound basis, fats pack almost twice the energy of carbohydrates, so for a mobile energy source fats are hard to beat. It is tempting to think that fat is always bad, but fat serves a number of important physiological functions. The accumulation of fat reserves for long-distance migration was one of the first forms of naturally occurring obesity to be investigated scientifically. Animals accumulate lipids in adipose tissue in order to sustain themselves over long annual migrations. Animals adjust their lipid stores to routes, weather conditions, and their own abilities in order to maximize the probability that they will be successful.

But how is excess glucose stored in adipose tissue? Fat is a connective tissue that has become specialized for storing nutrients. Adipocytes (fat cells) predominate in the tissue and are made up almost entirely of pure liquid fat. Mature adipocytes are among the largest cells in the body and are not capable of cell division. As they take up or release fat, they grow and shrink in size. Adipocytes account for about 90 percent of the mass of adipose tissue, and they are packed closely together, appearing almost like chicken wire when viewed under a microscope. Scientists have determined that there is a statistically significant relationship between body size and fat stores in mammals over an incredible range of body sizes, varying from ten to ten million grams.[58] Using comparative data from other mammals, based on our body size, we have about ten times as many adipocytes as expected.

This simply means that, because we have more adipocytes, the storage cells can be more numerous and can also have greater storage capacity. Our fat stores constitute a significant proportion of our body weight (15 percent in men and 22 percent in women, under normal conditions), ranging from 10 percent in the very lean to over 35 percent in the extremely obese.[59]

The obvious question is why should such a condition evolve? As in most of the explanations using evolutionary theory, we are not able to conduct the prospective experiment. We must rely on comparative data and post hoc experiments, at best. That is not to say that evolutionary explanations are meaningless but that they are based on a different kind of scientific reasoning than is familiar in the experimental scientific paradigm.

In a study of the role of diet and cardiovascular disease in captive crab-eating monkeys,[60] researchers unexpectedly found that animals fed diets high in fat and high in cholesterol were more sociable than monkeys fed a nutritionally adequate but low-fat diet. The researchers concluded that a low-fat diet promoted antisocial behavior by affecting lipids in the brain and altering the production of the neurochemical serotonin.[61] Serotonin has been strongly implicated in the mediation of aggression in both human and nonhuman primates.[62]

Fat is also necessary for the expression of a wide array of other social behaviors. Energy to engage in a variety of activities requires energy stores. Mammals spend a considerable amount of time simply being social. For nonhuman primates, social activities may take the form of grooming a favored partner, playing with another juvenile, or taking care of infants. Humans carry this trend toward elaboration of social interactions to the extreme, expending vast quantities of energy on social activities with no purpose other than simply being social.

In most animals, fat is necessary in one way or another for mating. Nutrient storage is necessary to attract mates, produce eggs or sperm, gestate fetuses, nourish infants, and offer parental care to immature offspring. Fatter individuals may be better able to defend territories, may be more successful foragers, may be better at incubating eggs, may be better at defending mates or offering high-quality parental care. Mothers rely on the transfer of lipids to offspring in most vertebrates as well as a number of other animals including insects, and fatty acids in a mother's milk are one of the main mechanisms of energy transfer.

One of the most utilitarian and typically overlooked functions of adipose tissue is restricted to humans. We, along with most of the nonhuman primates, have tactile pads under the tips of our fingers and toes, but in addition we have extensive padding under our heels. It may be that this padding is linked to our uniquely human mode of locomotion, bipedalism. The fatty components of the heel pad are tightly encased in cartilaginous tissue that is connected to both the heel and the skin. The necessity of this reinforcement is due to the high loading forces during the heel-strike portion of normal bipedal locomotion.[63]

Men and women differ in their average height and weight, with men being 8 percent taller than women and having about 20 percent greater body mass. However, one of the most dramatic morphological differences between human males and females lies in the soft tissue, the subcutaneous body fat. Adipose tissue can develop almost anywhere there is loose connective tissue but usually develops in subcutaneous tissue. (Fat also accumulates in bone marrow, around the kidneys, behind the eyeballs, and on the abdomen and hips.)

Women have approximately 19 percent more overall body fat than men. Not only do women have more; it is also distributed entirely differently than in men. In men, as in other primates, adipose tissue is deposited in the highly expandable "paunch" depot. Arising from the midline, at its maximum expression it can reach from the pubis to the neck, covering much of the belly and chest. For women, adipose tissue is deposited in the breasts.[64] Differences between men and women in the fatness of the limbs are much more pronounced than in the trunk. Fat differences in the trunk are about 5.7 percent, with a 35.4 percent difference in the arms and 46.7 percent difference in the legs and thighs.[65]

DIETING AND FAT

There is a significant difference in the use of the term diet when you are talking to a nutritionist compared to someone on the street. The common idea is that dieting is simply an attempt to lose weight, while dieting to a nutritionist means regulating the intake of all nutrients for good health. So what happens when you try to lose weight? One of the major observations of nutritionists is that body weight tends to be stable over time despite wide fluctuations in food intake. For most individuals, body weight and fat mass are maintained over years and even decades.[66] Overeating stimulates heat loss through heat release from brown fat stores.[67] On the other hand, dieting sets off a cascade of physiological alarms that have been shaped by natural selection in the face of episodic food shortages in our evolutionary past,[68] independent of nutritional status.

Oddly enough, initial efforts to lose weight result in the reduction of the mobilization of fat stores. This decline in fat breakdown is in direct response to reduced secretion of the thyroid in an attempt to slow the overall metabolic rate. This physiological response makes weight loss difficult even for the most dedicated in the face of heroic limitations of caloric intake. We have evolved the ability to "save up" energy for the inevitable food shortages that characterized our evolutionary past.[69]

As previously noted, there is an epidemic of obesity in the United States among adults, and there is little to suggest that future generations will be much different if current dietary practices continue. So, what causes obesity? The answer is that researchers are not sure. There are several theories that warrant serious consideration, however. The "clean plate club" theory suggests that overeating behaviors beginning early in life increase the number of fat cells formed, and the more fat cells there are, the more fat that can be stored. Blood-borne nutrients and hormones may actually deliver chemical signals to the brain that stimulate overeating. The increase in adipose tissue mass due to early overeating calls for an increase in the amount of fat stored. When there are substantial numbers of incompletely filled fat cells, plasma levels of fatty acids and glycerol are reduced, leading to abnormal hunger.

Fat cells themselves may also be implicated. When yo-yo dieters lose weight, their metabolic rate slows sharply. Nevertheless, when they subsequently gain weight, their metabolic rate increases, just like a steam engine being stoked. Each successive cycle of weight loss occurs more slowly, and the inevitable weight gain occurs more quickly (sometimes at a rate of as much as three times).

It may be that, when confronted with feast and famine in rapid succession, the body becomes more efficient, and metabolic rates adjust. The "high efficiency" theory suggests that obese people are more fuel efficient and more effective fat storers than people at normal weight. Calorie for calorie, fat is a much more concentrated source of energy than are proteins or carbohydrates. Normal metabolic processes require burning twenty-three calories from protein as compared to three from fat. Fat is almost eight times more efficient as an energy source because of the way it is metabolized. Couple that with the observation that, in obese people, the adipocytes develop more alpha receptors (which favor fat accumulation) and more of the enzyme lipoprotein lipase, which off-loads circulating lipids in the blood to the fat cells. Also, the off-loading in the obese is extremely efficient. The net result is there may be little correlation between caloric intake and body weight. In some sense, the deck is stacked against the obese.[70]

Imagine that studying mouse blood might offer a clue to obesity. Researchers at the Rockefeller University and the Howard Hughes Medical Institute found a hormone in mouse and human blood that affects food intake and metabolism. The hormone was first discovered in mice that have a defective gene, called the obese gene (Ob), that results in the accumulation of excess body fat in the affected individuals. A defect in the Ob gene is implicated in the production of leptin, a hormone that is thought to play an important role in the regulation of food intake and metabolism. Mice with the defective gene do not produce leptin. When obese (Ob) mice were treated with leptin [71] for two weeks, they lost 30 percent of their body weight.[72] As you can imagine, the implications of this study were staggering. Researchers rushed to learn more about the precise mechanisms involved in leptin production, and in the five years following the initial discovery, over four hundred papers were published in scientific journals reporting experimental results with leptin in mice and humans.

More recent research has led to major developments in understanding how leptin works. Leptin is produced in several organs as well as white adipose tissue, including brown fat, the placenta, and fetal tissues (such as heart and bone/cartilage). Leptin is implicated in inhibiting food intake, stimulating and maintaining levels of energy expenditure, signaling to the reproductive system, and, as a "metabolic" hormone, influencing a range of processes (e.g., insulin secretion, lipolysis, sugar transport).[73] Research has shown that leptin concentrations tend to be correlated with the amount of body fat in humans; however, most recent research has failed to deliver the results of tremendous weight loss promised by that first study.[74] Therefore, it appears that what was touted as a cure for obesity is actually part of a much more complicated system of weight regulation than first thought.

Researchers at the University of Wisconsin have offered some intriguing ideas about the cause of obesity. Animals inoculated with a virus (Ad-36) showed an increase in adipose tissue and paradoxically low levels of serum cholesterol and triglycerides. While research has not been conducted on humans, these data suggest that the role of viral disease in human obesity must be considered.[75]

Recently, researchers at the St. Vincent de Paul Hospital in Paris have focused their attention on the genes that indirectly exert some control of insulin production and of an insulin-related growth factor. They found that children who inherited one particular genotype from their father were far more likely to become obese as an adult. This variant, called Type I, is more frequent in central European and North African populations, with up to two-thirds of children in these areas possessing the Type I genes. Researchers noted that obesity rates in these areas are much lower than predicted by the genetic data, so there must be other factors influencing the genes' expression.[76]

An interesting part of the diet and exercise craze is the underlying motivation. Are people acting rationally in their own self-interest to reduce their mortality risk? For many people a reduced-calorie diet and a program of moderate exercise are recipes for longevity and health. However, for a substantial number of people the health benefits of dieting and exercise are not increased longevity but rather more immediate: the benefits are measured in increased physical attractiveness and access to high-quality mating partners. For this portion of the population, thin is in and fat is out. Regardless of the motivation, we are a society fixated on fat.

DIETING IS BIG BUSINESS

It is difficult to know the underlying motivation for every person who diets, but it is well known that, in the United States, dieting is a multibillion-dollar-a-year industry. Americans try all sorts of diet plans, exercise regimes, medications, treatments, and elixirs in order to lose weight. For many, dieting does not refer to attempts to control fat, sugar, or cholesterol, or a specific medical regime. Instead, dieting simply refers to achieving and maintaining lower body weight. It is a concern for and about calories, rather than nutrition, that drives many dieters.

Thin is in, but what is the underlying reason? It is well known that in American culture thinness is valued and corpulence is denigrated. The lesson of thinness is taught early and very effectively. Society dictates that if one is to be successful, one must be thin, and the not so subtle underlying message is that fat people are failures.

At the same time, there are enormous efforts to get Americans to eat and eat and eat. The food industry spends vast sums of money on advertisements targeting all ages and ethnic groups, even the very young. In a study of commercials on Saturday morning cartoon programming, approximately 60 percent focused on food.[77] Americans are drawn by images of delicious foods but at the same time are told that the consequences of obesity are tantamount to genetic suicide.[78]

The contrast between the hedonistic attitude toward food presented daily in advertising and the emphasis on dieting and weight loss is quite dramatic. No wonder mental health is a major issue in the United States. Physical attractiveness is defined in terms of a lean physique for women of all ages and for most men as well.[79] Preoccupation with physical appearance seems to be learned early on for both males and females, but girls greet the physical changes associated with puberty differently. Young boys tend to view the changes in their bodies as a means of achieving control over their physical and social environment. Maturation means enhanced competitive abilities to most young men. Girls, on the other hand, have a different attitude about puberty. As girls mature, their bodies change, and they often become increasingly dissatisfied with their appearance, consistently expressing a desire to be thinner.[80] Between 50 and 75 percent of the normal-weight females in a survey conducted by *Psychology Today* perceived themselves to be overweight, whereas only about 25 percent of the normal-weight males considered themselves overweight.[81] Whether or not females are overweight, they often see themselves as overweight and show decreased body-image satisfaction, reduced levels of self-esteem, and generally poorer psychosocial well-being than males in general and than females who do not perceive themselves as overweight.[82] Excessive weight is not simply a health issue, but it is also viewed as a character flaw associated with self-indulgence and laziness.[83]

One of the consequences of this emphasis on food, coupled with an almost pathological concern with thinness, is the efflorescence of the diet food industry. The diet industry is big business in the United States, grossing over $30 billion in 1990. That year the food industry spent over $36 billion in advertising just to get Americans to eat. Today, "lite" foods account for $14 billion in annual sales. Revenues of commercial diet centers exceeded $2 billion in 1990, and the clientele were mostly women (85–90 percent).[84] In 2000, Americans spent about $30 billion on diet programs, diet foods, and weight loss plans.[85]

The cost of dieting depends on the type of diet therapy undertaken. In 1997, a study conducted in Boston found that low-calorie diet programs cost over $2,000 for twenty-one weeks.[86] The tragic part is the enormous waste of money, for most will regain all the weight lost within two years.

INEVITABLE FAILURE OF DIETS

Americans are not getting fatter because people are consciously trying to gain weight. In fact, estimates are that at any particular time 25 percent of American men and 39 percent of American women are trying to lose weight.[87] Given the extraordinary concern with weight loss in American society, it comes as no surprise to find that there are literally hundreds of weight loss programs available, and they come in a variety of types: (a) balanced diets with mild reduction in calories; (b) high-carbohydrate diets, which contain plenty of sweets to satisfy sugar cravings; (c) high water content to increase bulk; (d) low-carbohydrate diets, which cause incomplete oxidation of fats and the subsequent accumulation of ketone in the body; (e) high-protein diets, which tend to be monotonous and hard to maintain; (f) high-fat diets, which taste good and take longer to digest so people feel full longer; and (g) a host of seriously unbalanced and potentially unhealthy diets (Beverly Hills diet, Carbohydrate Craver's diet, the Drinking Man's diet, Zen macrobiotic diets).[88]

In spite of this astonishing variety, the expected outcome of these dieting attempts is not bright. Those who complete the suggested dieting regime lose approximately 10 percent of their body weight, only to regain two-thirds of it within one year and almost all of it within five years.[89] Long-term evaluation of diet programs paints a grim picture. In a study of 192 men and women three years after the completion of a commercial weight loss program, researchers found that there was no significant change in average weight from the start of the study, even though individuals initially lost weight while in the structured weight loss program. In fact, only 12 percent maintained a significant proportion of their weight loss once they completed the diet program, while 47 percent gained more weight than they lost during the diet.[90] The bottom line to these long-term studies of weight loss programs is that for most participants they simply do not work.

Obesity in the developed world is a major health concern, and the only way to deal effectively with it is likely not to be pleasant. As Melvin Konner, a biological anthropologist and physician, noted in *The Tangled Wing: Biological Constraints on the Human Spirit*, the only way to lose weight is to become accustomed to feeling hungry.[91] However, we do not like feeling hungry, and many of the readers of this book have never really been chronically hungry, unlike many people around the world. Consequently, we have tried to find alternatives that would allow us to lose weight yet not feel hungry. Of course, the alternative to feeling hungry is to maintain calorie intake levels while increasing one's level of physical activity.

In addition to offering specific nutritional guidelines, clinicians have offered suggestions about healthy levels of activity. Researchers have found that there is a dose-response relationship between the amount of exercise performed[92] and mortality from all causes (i.e., the more you exercise, the greater the health benefits, up to a point) as well as mortality from cardiovascular disease in middle-aged and elderly populations.[93]

EATING DISORDERS AND CULTURAL PERCEPTIONS

One of the most tragic aspects of this emphasis on thinness and dieting is the resulting increase in the numbers of individuals with eating disorders. The emphasis on thinness has negative consequences for both the obese and the thin. Anorexia nervosa and bulimia nervosa are eating disorders characterized not only by excessive dieting but also by self-induced vomiting and use of laxatives to reduce weight quickly. Individuals suffering from these disorders often experience an inverse relationship between the fear of weight gain and actual weight loss. That means individuals are caught in an increasingly tighter downward spiral.

Anorexia nervosa is a self-induced condition in which patients frequently exercise excessively and starve themselves. Bulimia (from the Greek "ox-hunger") sufferers eat huge quantities of food, ingesting as much as fifty-five thousand kcal in a single hour, and then they induce vomiting or take megadoses of laxatives. Surprisingly, bulimics often have a normal body weight and a generally healthy appearance, but most suffer from swollen salivary glands, pancreatitis, and liver and kidney problems. Recurrent vomiting damages the stomach and the esophagus. Stomach acid flushed upward through the mouth eventually erodes tooth enamel. Bulimia in the extreme empties the body of potassium, causing a severe electrolyte imbalance and subsequent cardiac failure, or massive hemorrhage of the stomach.[94]

Anorexia frequently affects adolescent and young adult women. Ninety to 95 percent of anorectics are young and female, and as many as four females per thousand, twelve to eighteen years of age, may develop the disorder.[95] Patients are typically from middle- or upper-class families. There are few reported cases among American blacks, Hispanics, and first- or second-generation ethnic immigrants,[96] but this is likely due to socioeconomic status rather than particular genetic characteristics.[97]

The etiology of anorexia is unclear, but research suggests that the disorder is the result of a complex set of biological, psychological, and cultural factors. In its early stages, sociocultural factors that associate thinness with female beauty and fitness seem to play predominant roles, leading certain individuals into a pattern of chronic dieting.[98] As the cycle develops, individuals feel increasingly anxious about their weight, in spite of Herculean efforts to control it.

Data on the costs of eating disorders are difficult to obtain, in part because individuals with eating disorders are not likely to report the condition, parents are unlikely to admit that they have children with the disorder (many are in denial), and physicians are not trained to recognize the less ob-vious symptoms. The long-term outlook for patients suffering from anorexia is mixed at best. On average, only half of the patients recover from the con-dition, while 30 percent retain partial symptoms, and in 20 percent the con-dition becomes chronic. Reported mortality varies from zero to 21 percent with a mean of 5 percent over the last ten years, which means about one out of twenty anorexics dies every year.[99] Treatment of eating disorders is ex-pensive, and in its extreme form requires hospitalization. In most states, there are inadequate resources to meet the demand for treatment.

CONCLUSIONS

This chapter started out discussing the complex interplay between our culture and obesity. From a discussion of the physiology of fat to the secrets of extraordinary french fries, I have tried to highlight how our culture, es-pecially American culture, places an exceptionally high value on thinness as a primary index of attractiveness. Paradoxically, the United States is one of the most obese nations in the world. Reconciling these two observations is an exercise in unraveling the complexity of the interaction of culture and bi-ology. We are biologically predisposed to gain weight readily and metabo-lize it slowly and efficiently. This is a highly successful strategy, strongly favored by natural selection if you live in an environment of rapid, episodic, and frequent shortages of food. But that is not the case in the United States today. Americans live in a time of unprecedented availability of food, such that they waste over 25 percent (ninety billion pounds) of the food available for human consumption.[100]

Moreover, American culture has completely assimilated the idea of fast food. Massive marketing campaigns have been so successful that, if the present trend continues, it will not be long until many traditional restau-rants are out of business, particularly ones that take time to prepare meals from scratch. Grocery stores will have fewer fresh products that require ex-tensive preparation. There will continue to be a small number of grocery stores that cater to the affluent as well as those who favor "natural" foods. For most Americans, however, fast-food manufacturers will further insin-uate themselves into everyday life by offering their products in the frozen-food section of the grocery store. It is possible to buy frozen cheeseburgers today, and it may not be long before almost anyone can get a Big Mac and fries prepackaged off the grocers' shelves.

In addition to marketing, our biology also plays a nontrivial role in the ac-ceptance of fast food. The evolved preferences for salt, fat, and sugar have made us vulnerable to excesses that were unheard of in our evolutionary past. Our fat storage ability has now become a liability.

So, what are we to do? I am not completely sure, but I think it is unlikely that Americans will stop eating fast food because of some admonition that I might offer. What we might be able to do is use our knowledge of our evolutionary history to develop weight maintenance programs that focus on the prevention of significant weight gain in adolescents and juveniles. We might also use this knowledge to develop long-term treatments that view obesity as a disease and not a character flaw.

5 | Depression, Antidepressants, and Evolution

> If there be a hell upon earth, it is to be found in a melancholy man's heart.
> —Robert Burton, *The Anatomy of Melancholy*, 1621

Depression is one of the most common and debilitating psychiatric disorders; in fact, as a disabling ailment, it is second only to heart disease.[1] It is also one of the oldest known ailments. During the fifth century, Hippocrates was concerned about melancholia, a condition that appears to resemble what we know today as depression. While he was unable to arrive at a specific cause for the malady, Hippocrates classified it as a major mental disease that arose from an accumulation of excess "black bile," one of the four humors (blood, phlegm, yellow bile, black bile).[2] In fact, the Greek word for black bile was *melancholia*. Unfortunately, Hippocrates was not clear on how to rid oneself of this noxious substance.

By the nineteenth century, depression was of major concern to medical practitioners all over the world. Depression, or melancholy, is a central theme of many works of art, literature, and music. "The valley of the shadow of death" depicts life's gloominess,[3] while "Il Penseroso," by Milton, celebrates the pleasures of melancholy and solitude. American blues were born in the sadness of Negro slaves in the South. The cartoon character Sad Sack, drawn by George Baker, was a hapless and helpless soldier, resigned to his fate.[4]

WHAT EXACTLY IS DEPRESSION?

In spite of extraordinary efforts to describe and characterize depression over several centuries, our understanding of it, including its causation and treatment, has proven elusive. Depression affects an astonishing number of people worldwide. Ten percent of the entire population may suffer from it.

In terms of Western medicine, depression is today defined as a mood disorder by the *DSM IV, Diagnostic and Statistical Manual of Mental Disorders*, the clinical standard for psychiatrists and psychologists, published by the American Psychiatric Association. There are implications to defining depression as a clinical condition, but more about that later. According to the American Psychiatric Association, a major depressive disorder is characterized by at least a two-week period during which the patient experiences both a loss of interest in all pleasurable activities as well as four of these five symptoms: (1) changes in appetite, sleep pattern, or activity level; (2) decreased energy; (3) feelings of guilt or worthlessness; (4) difficulty in thinking or making decisions; or (5) recurrent thoughts of suicide. As a medical disorder, depression is differentiated from the physiological and psychological effects of alcohol intoxication or cocaine withdrawal. Clinical depression is also distinct from bereavement over the loss of a loved one.[5]

Shakespeare was acutely aware of depression, as demonstrated by Hamlet's fateful words, "How weary, stale, flat, and unprofitable seem to me all the uses of this world."[6] In his book *Why Zebras Don't Get Ulcers,* Robert Sapolsky, a neuroscientist at Stanford University, defines depression as a "genetic/neurochemical disorder requiring a strong environmental trigger whose characteristic manifestation is an inability to appreciate sunsets."[7] The emphasis in these descriptions of depression is that clinical depression is not a simple case of the "blues" but rather a crippling disorder that, when left untreated, can result in increased vulnerability to a variety of different conditions and, in its most extreme manifestation, suicide. Transient, short-term depression is a normal part of the human experience and is associated with any of a variety of day-to-day events. Everyone feels depressed at one time or another, but only a small proportion of the population suffers from major depression, as defined by psychiatrists.

Depression is in all likelihood not a single illness but a cluster of disorders. As such, classification has been difficult, but some consistent differences in symptoms have been noted. There seem to be two general types of depression: melancholic and atypical. Endogenous, or melancholic, depression is the clearest type and accounts for 40 to 60 percent of those hospitalized for depression. A subtype of endogenous depression is reactive depression, characterized by a precipitating cause. Reactive depression is thought to result from a specific stress (loss of a job, a spouse, or a pet) and appears to be an extension of normal responses to stress. Stress tends to overcome typical coping mechanisms. Reactive depression seems to occur in individuals who have a predisposition toward depressive behavior; however, individuals may not present with any of the five classic symptoms of endogenous depression.

It is abundantly clear that the distinctions between these two types of depression are not absolute, and in reality they may simply be the extremes of a continuum of melancholic depressive illness. Those suffering from atypical depression can be differentiated from the sufferers of melancholic depression. About 15 percent of those hospitalized for depression exhibit symptoms that are the opposite of melancholic depression. Individuals with atypical depression overeat and gain weight, have long periods of sleep, feel worse, not better, in the evening, and have pronounced periods of anxiety.[8]

The causes of depression have not been identified precisely, but there are certainly some observations that are suggestive. Data have accumulated to suggest that there is a genetic component to depression.[9] While there is no doubt that environment plays an important role, there seems to be a significant genetic component as well. This raises some interesting questions. Is it possible that the underlying basis for depression is actually a biological capacity, which in order to be fully expressed must have certain environmental inputs? While feeling blue is a part of the psychological experiences of all humans on occasion, there exists a proportion of the population that suffers from a condition that, in its milder form, affects virtually everyone. If this is true, we must accept the possibility that depression is an evolved psychological trait. That is, in our evolutionary past, individuals who exhibited at least some of the symptoms of depression must have experienced some slight advantages over those who did not exhibit the trait. As strange as it sounds, individuals who were depressed at least some of the time had a fitness advantage compared with those who were perennially perky. This perspective allows us to ask the obvious question about the function of depression in an evolutionary context. That is, how could such a potentially debilitating psychological condition ever have arisen in the first place?

EVOLUTION AND DEPRESSION

Recently, there has been increased attention to the idea that depressive states are a part of the inherited behavioral inventory of humans.[10] Among a small group of psychiatrists and psychologists interested in the evolution of psychiatric disorders, it is thought that depression may have an evolutionary basis. However, there is considerably less agreement on what adaptive function it might have served in our evolutionary past, or if it continues to function the same way in modern human society. British psychiatrist John Price noted that, while not a specific symptom of depression as clinically defined, one of the outcomes of depression is reduced activity and a reduction in social interactions, particularly competitive ones. At first it may seem strange to think that reduction in performance could confer an advantage to an individual, but certainly there are times when it is in our own interest not to pursue interactions with a particular individual. It may be that depression is a strategy that, under certain local ecological conditions, could confer an advantage on an actor.[11]

When considering depression and its possible functions, scientists have looked to the study of animal behavior for insights. We know that lethal combat is rare among animals and most conflicts are resolved without the use of physical aggression. Conflict resolution has been a hot topic in animal behavior studies for several years, and there is a significant accumulation of ethnological data documenting conflict resolution behaviors in a wide array of animals. These behaviors may be as simple and unobtrusive as moving away from a preferred resting site or food source, or not pursuing a particular mate. Conflict resolution behavior can also be quite complex and dramatic in the form of ritualized displays. In some species, these display behaviors have evolved, often augmented by actual signals of physical abilities (big claws, larger antlers, larger body size), to settle conflicts with minimal physical injury to the contesting parties. In these ritualized contests there is typically a winner and loser, and they behave in consistent and predictable ways. Animal behaviorists, often with little experience, can easily determine the winner and loser because the behaviors are so clear and unambiguous.[12]

Winning is associated with a particular constellation of behavioral and physiological responses. Similarly, losing has its own behavioral and physiological dimensions. What is important in these ritualized aggressive encounters is that there are clear signals of submission or defeat. The signals obviously serve important functions, for they prevent the escalation of aggression to lethal levels. From an evolutionary perspective, it is easy to imagine natural selection favoring the evolution of such displays. Contests over resources could be settled without participants incurring the costs of physical injury. In the terms of a popular television sports program, animals as well as humans experience *the thrill of victory and the agony of defeat* (the line made famous by Jim McKay).[13]

Now I want to consider the agony of defeat in more detail. A suite of behaviors exists that characterize defeat in humans (lethargy, slumped posture, gaze aversion, etc.) as well as in animals (cowed posture, rapid retreat, etc.). Psychological changes also characterize defeat, and it may be that these changes in psychological state are important mediators of the observable behaviors. Depression may be a behavioral mediator that, under conditions of conflict, could have been favored by natural selection. Exhibiting behavior that is consistent with a depressed psychological state could serve a number of functions. Depression could

1. prevent an individual from attempting a comeback against clearly superior opponents and suffering significant physical injury;
2. induce behaviors that would signal "I give up" to opponents, thereby avoiding additional harm or injury;
3. induce behaviors that would signal "I am no threat" to a potential opponent, avoiding any type of competitive encounter;
4. act as a coping strategy to deal with subordinate status, a way of accepting the reality of defeat.

What these suggested functions have in common is that they all have to do with minimizing the costs of competitive encounters. For many of us, it may be better to live to fight another day than to incur serious injury, and depression may be an evolved psychological trait that helps us do that.

Depression and sadness may also be adaptive in a slightly different way. George Williams, an evolutionary biologist, and Randy Nesse, a practicing psychiatrist, wrote in their recent book, *Why We Get Sick*,[14] that depression and sadness often signal the loss of resources that are important in reproduction (wealth, power, status, health, and allies). Depression then becomes a signal to the afflicted individual that he or she has been doing something that resulted in this loss and that, by stopping a particular behavior, the individual can stop present losses as well as ward off future losses.

The experience of a loss should signal us to stop doing what we are doing. Even the thought of the pain associated with slamming one's fingers in a rapidly closing car door makes us more careful and less likely to engage in this behavior. So too does the psychological pain associated with depression reinforce certain behaviors. Avoidance of the pain of depression can be a strong motivating force in shaping human behavior.

Depression could also serve to delay making important life decisions. While it is true that some life decisions must be made in the blink of an eye (e.g., avoiding an oncoming truck on the highway, finding a tree to climb when being chased by a lion), others are best not done precipitously. A note to procrastinators: this does not mean that evolution is an excuse. Evolutionary ecologists and mathematical game theorists have long recognized the importance of not abandoning too quickly an enterprise into which one has invested a lot of time and resources.[15] One would hope that investors in the stock market would also understand this idea. On the other hand, individuals reach a point where abandoning an enterprise is the best course of action. Depression might serve to make us less likely to abandon enterprises, social relationships, or economic endeavors into which we have invested heavily, but clearly there are times when we have to cut our losses and move on to new opportunities. This reluctance brought on by depression could have served an important function in the evolving human psyche. It is well known in the clinical literature that depressed individuals often miraculously cure themselves when they abandon some long-sought goal and redirect themselves to other endeavors.

This evolutionary perspective suggests that, rather than being some horrible life-threatening disorder, depression may be in most individuals an evolved psychological coping mechanism. If this is true, should we be concerned about depression, or should we simply "Buck up, buckaroo"? I think that the answer is both yes and no. It is important not to use an evolutionary explanation as an excuse for a particular pattern of undesirable behavior (e.g., physical aggression is so deeply rooted in humans that it is inevitable we will engage in warfare; or stepchildren will be the target of abuse from stepfathers, so males should not be allowed to adopt offspring), but we can use this perspective to help us understand the condition or behavior.

To discount and minimize the psychological toll that severe depression has inflicted on millions of people is not helpful. In the same vein, however, the reality is that some depression is an inevitable part of life. Evolutionary hypotheses of depression emphasize the adaptive nature of particular behaviors in the face of defeat in social interactions, the loss of resources, and so forth. The evolutionary perspective does not suggest that for millions of people depression is an inevitable debilitating, chronic, and sometimes fatal psychological state. The interesting question is how to account for the differences between a good old-fashioned case of the blues and a serious clinical condition. The truth is that no one knows for sure. There is a growing body of knowledge suggesting that those who suffer from true clinical depression have some fundamental changes in the chemistry of their brains. These chemical imbalances, not some weakness of character, are the underlying basis for depression.

MENTAL ILLNESS AS A "CHARACTER FLAW"

While depression has long been of concern in the medical community, mental illness in general has suffered from the negative stereotypes ascribed to patients. There is a widespread, largely unstated idea that people suffering from mental illness have some sort of character flaw. If only they had a real sense of character, then they could "pull themselves up by their bootstraps." While the reasons for this view of mental illness are certainly complicated, it seems that they are reflections of a standard of behavior originally articulated in the sixteenth century by the French religious and social reformer John Calvin. The philosophical bases for this view of mankind are beyond the scope of this chapter, but the idea of rugged individualism and self-reliance seems to have been taken too far.

Little social stigma is associated with those who suffer from cancer, atherosclerosis, or diabetes, but not so for the sufferers of mental illness. Not only are there societal pressures against the recognition of psychiatric disorders as legitimate medical conditions: there are also significant hurdles imposed by the health care industry. The net result of these efforts is that, according to some authorities, only one in three people who are clinically depressed gets treatment. Depressed individuals often experience a disabling constellation of physical symptoms that may include chronic pain, incapacitating fatigue, and persistent headaches. In addition to the psychological symptoms, these physical manifestations often make it difficult for the suffering patient to get adequate treatment.[16] Often the physical symptoms of depression are the ones that are treated, without attention to their underlying cause. Individuals who experience major depression suffer as much as those who have more traditional diseases such as high blood pressure, diabetes, atherosclerosis, or even hives.

HOW WIDESPREAD IS DEPRESSION?

The exact prevalence of depression is unknown, but it has been estimated that, during one's entire lifetime, 15 percent of the men in the United States and 24 percent of the women will experience a major depressive disorder.[17] Others suggest that 6 to 22 percent of patients in primary care suffer from a depressive disorder.[18] Surveys of health care clinics in sub-Saharan Africa and Latin America report that, in one-fifth to one-third of the cases, depression is the primary or secondary reason for seeking health care.[19] Recent studies have shown that rates of depression seem to increase and the age of onset tends to decrease with rapid modernization in society, as seen in Puerto Rico, Taiwan, and Lebanon.[20] A survey of mental disorders suggests that the lifetime prevalence of depression may be as high as 20 percent worldwide. In the same survey, depression was found to be the most frequently encountered mental disorder.[21] If these data are true, we are experiencing a worldwide epidemic of depression.

Is it possible that one of the unintended consequences of worldwide economic development (or exploitation, depending on your perspective) and the spread of technology is wholesale depression? At one level, this explanation for the prevalence of depression makes sense. The human brain, like other aspects of our anatomy and physiology, is the result of millions of years of evolution. Likewise, many of our behavior patterns are also the products of this long evolutionary history. But the gristmill of natural selection typically grinds exceedingly slowly. It may very well be that individuals who experienced mild depression could have accrued some evolutionary advantages in a society characterized by small groups of individuals who were likely related and had persistent and intimate face-to-face contact as well as mutual reliance and cooperation for survival. However, life is very different now from that of 150,000 or so years ago when our species arose. Modern Western society is being exported around the world today. There are few places in the world not affected by television. The adaptations that served our ancestors well since Pleistocene times may not be useful now. Is it possible that the discontinuity between our evolutionary past and our modern culture is contributing to the outbreak of depression? Many researchers say yes. But is there an alternative explanation for the excessive levels of depression seen today?

To be sure, serious depression is nothing to take lightly, but Edward Shorter, a historian with an interest in medicine, in his book *A History of Psychiatry: From the Era of the Asylum to the Age of Prozac*, offers an interesting perspective on depression in the United States.[22] He wrote that one of the problems facing psychiatry is the tendency of people to seek alternative psychological or sociological solutions to problems that were previously the province of psychiatry. Psychiatry began to lose out to nonmedical forms of counseling, such as clinical psychology and social work. Postmodern introspection has taken over the American psyche, and the boundaries of traditional psychiatry have been invaded by nonclinical approaches and, to some extent, nonscientific thinking.

Increasingly, Americans in record numbers are seeking help for mental problems. Two things seem to be at work here. First, the definition of what people consider mental illness has changed considerably. Boundaries have been blurred and definitions expanded. Psychiatry is responsible in no small part for this trend, because as the increase in the number of psychopathologies has grown, so has the need for psychiatrists.

Adolescence was a primary target for increasing the market share of mental health for psychiatry. The phrase "boys will be boys" applied to the devilish pranks of young males, which had previously been ascribed to nothing more than the exuberance of youth, has become pathological. In the current climate of labeling, almost any behavior of adolescents can be pathological; Huck Finn and Tom Sawyer could be diagnosed with minimal cerebral dysfunction. This diagnosis has lost favor, being replaced by the view that overactive young males suffer from "hyperkinetic reactivity of childhood or adolescence." Depression was another one of those conditions that was an easy target for expansion. In popular terms, the definition of depression has come to be equivalent to dysphoria (unhappiness) combined with the loss of appetite and insomnia.

The American Psychiatric Association launched a program in 1991, in the context of its "Mental Illness Awareness Week," called "National Depression Screening Day." The program encouraged nonpsychiatrists, particularly general practitioners, to become more aware of depression and diagnose it in their patients. Now let me not sound too cynical about all this, since we know that a missed diagnosis can result in suicide. However, the net result of the increasing awareness was exactly what the psychiatric community may have been looking for: an increasing patient base. Whether depression is actually on the increase or there is an increasing awareness of it, depression now accounts for two-thirds of patient visits to psychiatrists.[23] Depression is the most common mental health disorder in the United States, with over seventeen million people suffering from it. The broadening of the definition of depression by psychiatrists was the broadening of the definition of disease. Some psychiatrists became increasingly aware that what they were treating was not major depression but rather individuals who were profoundly dissatisfied with the quality of their life. According to some, the definition of mental illness has become so diluted that it is no longer meaningful.

DRUGS AND DEPRESSION

We have long had a fascination with mind-altering drugs. In art as well as literature, we have been fascinated with the possibility that drugs could alter behavior on a large scale. Recall the fictional drug "soma" in Aldous Huxley's *Brave New World*, Thomas de Quincy's *Confessions of an English Opium Eater*, Sherlock Holmes's addiction to cocaine, and Stan Emery in *The Manhattan Transfer*, who saw drinking as the only means to adjust to the world. Popular music has also been heavily influenced by the effects of widespread drug use. Songs by many groups extol the virtues of drug use ("Cocaine" by Eric Clapton, "White Rabbit" by the Jefferson Airplane, "Sky

Pilot" by the Animals, "Journey to the Center of the Mind" by the Amboy Dukes, and "Eight Miles High" by the Byrds, for example).

The ancient Greek poet Homer was one of the first to be concerned about the mood-altering effects of certain substances. In the *Iliad*, Homer tells the story of Helen of Troy and her kidnapping by the Trojans, which incites the Trojan War. After the war, Helen and her husband, Menelaus, while traveling (to/through) Egypt encounter the drug dealer Polydamna, who sells her a drug, *nepenthe*, which is supposed to cure sadness. At a party years later, Helen's husband and their guests begin to lament the loss one of Menelaus's closest friends, Odysseus. Helen mixes nepenthe with the wine being served. While the conversation about Odysseus continues, the grief suffered by the guests abates. Because cannabis was unknown in Egypt until over a thousand years after Homer, it is likely that the miracle antidepressant was opium. Albrecht Dürer, the German Renaissance artist, created an engraving in 1514 that portrayed "the peculiar wretchedness of the drug addict," titled *Melancholia*. These examples provide only a brief glimpse of the centrality of drugs in the arts.[24]

Today, fiction has become reality. The discovery of antidepressant drugs is one of those interesting stories in science and in some ways almost seems as if it were taken from the lyrics of a song. In 1952, a drug called iproniazid was introduced to treat tuberculosis, and it met with astonishing results. Not only was there remarkable improvement in affected individuals, but the drug also stimulated appetites, increased energy, and induced euphoria. Newspapers in 1953 ran a photograph from the Associated Press that showed patients in the Sea View Sanatorium in New York, dancing in the corridors of the hospital. The caption of the photograph read, "A few months ago, only the sound of TB victims coughing their lives away could be heard here."

The celebration of those patients at Sea View was not due completely to the therapeutic effects of the drug. Nathan Kline and John C. Saunders were the first physicians to become interested in the effects of iproniazid on behavior and mood. They referred to iproniazid as a "psychic energizer" and were convinced that it might be of help in treating depression. In 1957, the *New York Times* reported on the proceedings of a research conference and included news of the preliminary success of iproniazid in treating depression. The results were encouraging and were confirmed in other studies in the United States as well as in Europe. More than 400,000 depressed patients were treated in the first year. Unfortunately, there were a number of noxious side effects, not the least of which was the development of jaundice in 127 cases. The toxicity of iproniazid rendered its use problematic, but it was quickly replaced with other drugs with similar modes of action.

At the time of the discovery of iproniazid for depression, Swiss psychiatrist Ronald Kuhn was interested in the antidepressant properties of opium. Opium caused some of the symptoms of depression in healthy subjects, but it also seemed to decrease them in those already depressed. Although Kuhn was developing a theory of depression based on the action of opium, it is likely that the successes he witnessed may have been due to the general mood-altering properties of opium that alleviated a variety of symptoms, not just depression. The hallmark of opium was that it restored energy in affected individuals without being a stimulant. Kuhn argued that an effective antidepressant would be one that acted in a specific manner against melancholy and was not just a general stimulant.

Kuhn began looking for such a drug among a class of compounds widely used to treat allergies, the antihistamines. While antihistamines did have the effect of reducing the symptoms of allergies, they also had a sedating effect. Kuhn was interested in antihistamines because the first modern psychotherapeutic medication, chlorpromazine, was an antihistamine. Chlorpromazine (Thorazine™) was introduced in 1952 with remarkable success in treating schizophrenia. Kuhn became interested in antihistamines that had a similar chemical structure to chlorpromazine. Less than six months after the introduction of iproniazid for the treatment of depression, Kuhn announced that he had found an experimental drug, called G22355, that sedated normal people but relieved the depression of affected individuals. It was named imipramine (Tofranil™) and was the first nonstimulating antidepressant. While the initial response to imipramine was mixed, it turned out not to have the noxious side effects of iproniazid and ultimately enjoyed widespread use in the therapeutic setting.[25]

Soon after their introduction, both iproniazid and imipramine were shown to affect how nerve cells communicated. The theory of mood regulation at that time was the amine hypothesis. Briefly, this hypothesis suggested that amines, naturally occurring neurotransmitters, were not properly regulated in those suffering from depression. Like the diabetic who suffers from a shortage of insulin or the sufferer of Graves' disease (hyperthyroidism or an overfunctioning thyroid), individuals with amines that were out of whack could suffer from mental disorders. Those with an excess of amines were thought to suffer from mania, while those with a deficiency suffered from depression. It was suggested that nerve cells communicate by secretion of a substance into the space between the cells.

Then, the message is completed in a two-stage process in which the amines are taken back into the transmitting cell and inactivated by "janitorial" enzymes. Imipramine slows the reuptake of amines by the transmitting cell, thus leaving a high concentration of amines at the synapse for a longer period. Iproniazid inhibits the production of the "janitorial" enzyme, making more of the amines available in the cell.

At the time, some data called the amine hypothesis into question, or at least posed problems for the excess/deficiency aspects of the hypothesis. Specifically, there is a time lag in the effects of any antidepressant. Antidepressants have an effect at the neurochemical level in a matter of minutes or at least hours after administration, but depressed patients report no mood change for as many as four weeks. The amine hypothesis does not offer an explanation for this curious finding. According to the amine hypothesis, any drug that depletes amines in the brain should cause depression. It was well known at the time imipramine was discovered that certain drugs, namely reserpine (a drug used to lower blood pressure) lowered amine levels, but only caused depression in about 20 percent of the patients using it. Clearly, depletion of amines was not the only cause of depression.

In the wake of the discovery of imipramine, a whole host of drugs were developed that were structurally similar to imipramine, and all had the same general effects. Now there are several drugs, all similar in chemical structure, called tricyclic antidepressants (named for their three-ring molecular structure) that are effective in the treatment of the symptoms of depression. Imipramine and the other tricyclics are effective antidepressants in 60 to 70 percent of those suffering from classic endogenous depression. However, there are some serious limitations to the use of the tricyclics.

Remember that when Ronald Kuhn discovered imipramine he was investigating various antihistamines. One of the effects of antihistamines is to interfere with another neurotransmitter, acetylcholine. When acetylcholine nerve transmission is pharmacologically interrupted, one of the consequences is to activate the body's fight-or-flight response. Because of its biochemical similarities to some of the antihistamines, imipramine was found to cause some of the same side effects: sweating, heart palpitations, dry mouth, constipation, and urinary retention. Interestingly, iproniazid and its relatives also induce the fight-or-flight response, but not as severely as imipramine. This difference alone would have tipped the balance in favor of prescribing iproniazid for certain kinds of depression, but it was also found to have the particularly undesirable property of elevating blood pressure. The net result was that for several years the drug of choice was imipramine.

Some compounds similar to iproniazid were also found to be important in the treatment of depression. Remember the janitorial action of iproniazid in amine regulation. Monoamine oxidase (MAO) is the janitorial enzyme that oxidizes certain amines. MAO inhibitors effectively increase the amount of amines in the brain because of their blocking the janitorial effects of MAO. Monoamine oxidase inhibitors remained the major antidepressants for several years, but an increasing body of data showed that their interactions with other drugs and with amines in a variety of foods can lead to hypertension. This resulted in their fall from favor. A rash of deaths from cerebral hemorrhage among patients taking MAO inhibitors, as well as severe headaches experienced by some patients due to extremely high blood pressure, led to a dramatic decline in their popularity as an antidepressant.

More recently, the recognition of one particular neurotransmitter has been especially important in the development of more effective antidepressants with fewer side effects. It is serotonin (5-HT), a primary inhibitory neurotransmitter in the brain. Serotonin is intimately involved in a variety of neuropsychiatric activities including appetite control, regulation of the sleep-wake cycle, circadian cycles, as well as brain development and maturation. Disruption of the regulation of serotonin in the brain is implicated in a number of psychiatric disorders (depression, anxiety, obsessive-compulsive disorder, eating disorders, seasonal affective disorder, alcoholism, schizophrenia, Alzheimer's disease, autism, attention deficit disorder, Tourette's syndrome, and sociopathy).[26] The recognition of the importance of serotonin has led researchers to find medications that might mediate disruption of the serotonin regulating system.

One of the first drugs marketed that directly affected serotonin was trazodone (Desyrel™). However, as with the earlier tricyclics, trazodone has proven to have unpleasant side effects. Trazodone was extremely sedating, and often, before therapeutically effective doses could be administered, patients became tired and dizzy. This was the state of clinical treatment for depression in the early 1980s and was the milieu into which the new class of antidepressants was introduced.

As with other psychoactive drugs, serendipity played a part in the development of the first serotonin reuptake inhibitor, fluoxetine hydrochloride (Prozac™). The story of fluoxetine begins in the 1960s with an organic chemist, Brian Molloy, who worked for the pharmaceutical giant Eli Lilly. Molloy was interested in producing cardiac drugs, and he was led to acetylcholine because of its cardiovascular effects. Another chemist came to work at Lilly, Ray Fuller, who persuaded Malloy to abandon his research on cardiac drugs and begin work on a drug that would affect amines, but not acetylcholine, and consequently the new drug would not have the undesirable side effects of dry mouth, tachycardia, and urine retention.

The third player in the story is Robert Rathburn, who was interested in the effects of antihistamines on hypothermia in rats. Hypothermia was induced by administration of apomorphine. Rathburn found that certain antihistamines would block the apomorphine-induced hypothermia, but not antihistamines that also affected acetylcholine. Malloy decided to use Rathburn's model as a way of testing new antihistamines to search for one that did not affect acetylcholine. He chose one of the most common antihistamines, diphenhydramine hydrochloride (Benadryl™), as a place to start.

The final character in this story is David Wong, another chemist at Lilly. Wong's research focused on mechanisms of antibiotic action, but he had become interested in the neurochemistry of mental disorders. He became convinced that serotonin was the neurotransmitter that might yield new insights. The collaboration between Malloy and Wong actually came about because they both attended a lecture by Solomon Snyder of Johns Hopkins University. Snyder's talk reviewed his work on isolating neurons that were involved with biogenic amines.

It is useful to take a quick diversion to review some fundamentals of neuroanatomy in order to understand more fully what these scientists were up to. Nerve cells, or neurons, communicate with each other at special junctions called synapses (from the Greek word that means to "fasten together"). A nerve impulse runs down one of the arms of the nerve cell, called the axon, and arrives at the end of one or more endfeet (or terminal buttons, in more technical terms). When the nerve impulse arrives at the endfoot of an axon, it triggers the release of a chemical called a neurotransmitter into the synapse from tiny membranous spheres called vesicles. The area between the presynaptic membrane of the endfoot of the transmitting cell and the postsynaptic membrane of the receiving cell is called the synaptic cleft. This gap between the two neurons is where the action really takes place. The gap is about twenty to thirty nanometers wide and contains a fluid, which is poorly understood chemically.[27] The synaptic cleft is also made up of tiny strands that hold the pre- and postsynaptic membranes close to each other. These filaments, called the synaptic web, are exceedingly strong. In fact, they are so strong that, when brain tissue is broken up for analysis, the presynaptic membrane, the synaptic web, and the postsynaptic membrane often stay together. The whole complex—the endfoot, the synaptic web, and the postsynaptic membrane—is called a synaptosome, and it was the synaptosome involved in biogenic amine transmission that was the focus of Solomon Snyder's lecture. Remember, the lecture was the reason that Malloy and Wong began to entertain the possibility of working together.

In his lecture, Snyder described the procedures he had developed that would allow the production of a brain extract (obtained by grinding the brain and separating the synaptosomes) that acted like living nerve cells. You could expose the extract to a neurotransmitter and see how much was absorbed. Not only did Snyder lecture on synaptosomes, but later he also instructed Wong and Molloy on what neurochemists call "binding and grinding."

Using the techniques developed by Snyder, Wong was able to study the compounds that Malloy had developed as possible antidepressants. In fact, compounds that passed the apomorphine test were not interesting; it was the ones that *failed* that turned out to be most interesting. One of these compounds, 82616, blocked the uptake of serotonin and little else. The drug was tested in several different labs and was found to selectively block the reuptake of serotonin into the transmitting nerve cells. Experimental drug 82616 was fluoxetine oxalate. Soon it was found that it was easier to work with a similar compound that had the same properties as 82616. That was fluoxetine hydrochloride, the drug that would later be marketed as Prozac™. Wong and Malloy published the result of this work in June 1974, in the scientific journal *Life Sciences*. Years passed before Prozac was ready to be marketed, and finally, in December 1987, it was released.[28]

The best way to describe the action of fluoxetine compared with the other antidepressants being used in the late 1980s is by analogy—comparing, say, the destructive firepower of a Scud missile in its indiscriminate lethality to the specificity and precision of a laser-guided heat-seeking Maverick missile. Although more specific in their mode of action than the tricyclics, the new group of antidepressants acting on the reuptake of serotonin, called selective serotonin reuptake inhibitors (SSRIs), was not as precise as clinicians would have liked.

Initially, Prozac was met with an unremarkable degree of enthusiasm. Clinical studies had shown that it was only marginally less effective than the current gold standard for treatment of depression, imipramine. The major difference was the lack of unpleasant side effects and the ease of dosing (no need for titration and a once-a-day schedule). The long half-life of the SSRI drugs provided an additional advantage in protecting against breakthrough symptoms or relapse if a dose was missed.

Side effects of SSRIs occurred in a small percentage of patients (roughly 1 to 15 percent, depending on the side effect, the drug, and the dosage) in

clinical trials. They included nausea, somnolence, agitation, tremor, dizziness, nervousness, decreased appetite, insomnia, sweating, and, particularly in males, sexual dysfunction.[29]

The reported lack of unpleasant side effects in a majority of patients changed the way that mental illness is treated in Western medicine. Now, for the first time, psychiatrists had a drug that would treat depression but did not require a specific diet, did not increase risk of strokes, and did not leave patients feeling drugged. All of these were factors that interfered with the psychotherapeutic relationship between doctor and patient.

The SSRIs have been recognized as effective treatment for those suffering from mild to moderate depression, but their usefulness in severe depression is unclear.[30] Today, there are a number of SSRIs available, and their precise mode of action varies, but the general pharmacologic action is on serotonin and its receptors. It is also becoming clear that, rather than simply treating depression, SSRIs may be useful in treating all sorts of conditions involving mood disorders, anxiety disorders, obsessive-compulsive disorder, post-traumatic stress disorder, eating disorders, migraine headaches, schizophrenia, seasonal affective disorder, Alzheimer's disease, Tourette's syndrome, borderline personality disorder, and suicidal tendencies.[31] In light of this wide array of conditions on which SSRIs have some effect, some caution might be in order, given our limited overall knowledge of how the brain actually works.[32]

DRUGS, MONEY, AND BIG BUSINESS

In general, psychotropic drugs (drugs that act on the brain) are among the most widely prescribed medications in the United States, with sales exceeding $23 billion in 2000 and expected to reach $42 billion by 2005. Of the psychotropic drugs, antidepressants are the best-selling type of prescription medication. Retail sales of prescription antidepressants totaled $10.4 billion in 2000, up almost 21 percent over 1999. Antidepressant drugs accounted for an unprecedented 7.8 percent of total prescription drug sales in 2000. Of the top ten best sellers in 2000, three were antidepressants, followed by two antiulcerants, two cholesterol reducers, one antiarthritic, one oral antihistamine, and one oral diabetes drug. The number of prescriptions for all antidepressants increased from 40 million in 1988 to over 120 million in 1998.[33]

Why have Americans experienced such a dramatic rise in the use of antidepressants? Is it possible that the number of depressed people in the United States is increasing? If so, is this increase out of proportion with annual population growth?

While data are imprecise, at best, it is clear that the use of antidepressants is increasing at a rate that far outstrips the average population increase of about 0.9 percent per year. Even if the increase in actual population cannot account for the increase in antidepressant use, it is true that the demographic composition of the population is changing, as the baby boomers near retirement. Older patients tend to visit physicians more than younger people, and, when they do visit a physician, they are more likely to leave with more prescriptions.[34]

Some part of this dramatic increase in the use of antidepressants is likely due to the numbers of people seeking care because of increased knowledge of depression through the efforts of recent depression awareness programs, best-selling books, or other media attention. Alternatively, diagnostic patterns have changed, with individuals diagnosed more frequently for depression than before. It is also conceivable that the increase is due in large part to the availability of a number of effective, well-tolerated medications that encourage physicians to evaluate their patients more carefully for depression. It is interesting that most prescriptions for antidepressants are written not by psychiatrists but rather by internists and pediatricians.

A final possibility is that the actual rate of depression is increasing, not only in the United States but across the world. As indicated in some of the cross-cultural data, the number of people worldwide diagnosed as suffering from depression is increasing. It is difficult to get a good estimate of the true prevalence of depression, however, for a variety of reasons, including: (1) the definition of normal versus pathological mood is not clear; (2) case sampling is difficult; (3) non-Western depression is often not recognized; and (4) there are no tests of diagnostic reliability in a cross-cultural setting. What is interesting, of course, is that the cultural manifestations of depression vary considerably, based on ethnographic studies of particular groups.

Culturally equivalent terms for depression are lacking in many non-Western societies (Nigerians, Chinese, Eskimos, and Malaysians).[35] That does not mean that individuals living in those cultures are not depressed; it simply means that individual interpretation of the depressive experience is subject to considerable cultural variation. Research designed to characterize depression in a cross-cultural context is needed in order to appreciate the magnitude of depression as a pancultural mental state. If depression is on the increase, it becomes even more urgent to understand more fully the factors that precipitate it and develop strategies to minimize its deleterious effects on overall health and well-being.

One of the most disturbing trends in the United States is the increasing use of psychotropic medications on adolescents and children and, in particular, the trends in prescribing antidepressants. A recent study reported that patients under eighteen were over six times more likely to be prescribed an antidepressant in 1994 than in 1985.[36] These findings controlled for a number of confounding variables including sex, race, and diagnosis. Approximately 30 percent of the adolescent or child psychiatric visits in a 1993–1994 survey included an antidepressant medication.

Of nearly six hundred family physicians and pediatricians who responded to a 1999 University of North Carolina survey, 72 percent said they had prescribed antidepressants to children under the age of eighteen. Only 16 percent of the doctors said they felt "comfortable" doing so, and just 8 percent said they have adequate training to treat childhood depression.[37] This is in marked contrast to current prescribing practices in Great Britain, where the average child psychiatrist started antidepressive therapy in only one or two children each year.[38]

This difference in prescribing trends suggests two vastly different approaches to the treatment of psychiatric disorders in children. Studies of the treatment of depression in children have failed to produce clear and unambiguous results favoring the use of antidepressants. In a recently published meta-analysis of thirteen published, double-blind, placebo-controlled studies of antidepressants in children, psychiatrists at the State University of New York Health Center concluded that, "except for two possible borderline instances of the drug exceeding the placebo in efficacy, there has been a consistent failure to demonstrate any advantage over placebo."[39] The authors conclude that, in spite of the lack of data supporting the efficacy of treating children and adolescents with antidepressants, clinicians have come to use antidepressant therapy on such a wide scale that it is a standard of practice in many areas.

This widespread use of antidepressant therapy is disturbing for a number of reasons, not the least of which is the lack of proven effectiveness. More troubling is the paucity of knowledge about the effects of psychotropic drugs in general, and antidepressants in particular, on child and adolescent neurobehavioral development. There are studies that have demonstrated the influence of maturational effects on drug action. In a classic study of the action of amphetamine on mood, investigators found that the drug induced feelings of euphoria in adults and adolescents, but it did not do so in prepubertal children.[40] Epidemiological studies show that the rate of affective disorders nearly doubles from prepuberty to adolescence and the sex ratio changes from almost equal or a slight male preponderance among prepubertals to a pattern of female preponderance by adolescence. The rapidly changing hormonal status of the brains of children and adolescents, the incomplete maturation of neurotransmitter systems involved in the control of mood, as well as differences between the ways drugs are metabolized in children, adolescents, and adults should alert us to the problems that are present when we try to treat children with drugs designed for adults.

Use of antidepressants by pregnant and nursing women is also troublesome. The SSRIs are rated pregnancy category B drugs, which means that there were no harmful effects found in animal studies. However, there are no reliable data on safety in humans during pregnancy. A recent epidemiological analysis of studies of nursing infants of mothers taking antidepressants reported adverse effects on infants whose mothers were treated with fluoxetine.[41]

In considering the possible long-term adverse effect of antidepressants on neural development and maturation, it is also useful to take a careful look at the history of other highly touted "magical drugs." Of course, one of the best-known cases of misplaced enthusiasm for a psychotropic drug was Sigmund Freud's advocacy for cocaine. Early in his career, Freud saw no possible risk of addiction to cocaine, and even later in his career, when confronted with patients that he had addicted to cocaine, Freud responded that the addiction was due to some other personality problem and not the cocaine. In the wake of the widespread therapeutic use of cocaine followed the practice of prescribing heroin to treat cocaine addiction.

DANGERS OF PSYCHOACTIVE DRUGS

It took over twenty-five years after their introduction to establish the addictive potential of the benzodiazepines. Librium, a powerful benzodiazepine, was the focus of an ad in the *British Medical Journal* (1 March 1969) that proclaimed, "Whatever the diagnosis—Librium." Librium and Valium are both widely prescribed today in spite of their well-documented addictive potential. In some ways, there are parallels with the popular view of antidepressants, particularly the SSRIs.

The view of fluoxetine as a magic bullet for treatment of a variety of psychotropic conditions is the result of several factors. One factor in this headlong rush to Prozac and the SSRIs has been the best-selling book *Listening to Prozac*. In this book, psychiatrist Peter Kramer chronicles his experiences with the drug and concludes that it has the potential to change the lives of patients. Moreover, he argues that Prozac has the possibility of producing, in otherwise healthy users, increased self-assurance, greater hope for professional, social, and personal growth, and a love of life. Kramer coined the term "cosmetic psychopharmacology" to describe the effects of this wonder drug on behavior.

In spite of such extraordinary claims for Prozac in particular and the SSRIs more generally, there are some unsettling observations that should make potential users cautious. Clinicians, as well as the drug manufacturers, claim that antidepressants are not addictive and that people are mistaken if they think that antidepressants can induce dependence.[42] Such claims fly in the face of the recommendation that these drugs should not be abruptly terminated but instead a reduction of dosage should be carried out over the course of several weeks.

It seems that what the manufacturers are saying is that the drugs do not lead to overt drug-seeking behavior and hence they have little abuse potential. While on the surface this may be true, the claims of nonaddictiveness do not address the problem of the withdrawal symptoms from the SSRIs. Moreover, it is well known that some SSRIs have an extremely slow clearance rate, so that the withdrawal symptoms tend to peak long after the drugs have stopped being taken. The problem is that withdrawal symptoms may not be recognized for what they are, particularly in individuals with relatively little experience in dealing with psychotropic medications.[43]

In a clinical trial of sertaline, in addition to the side effects that seem to be common to all the SSRIs, there was a report of a number of side effects upon discontinuance of the drug for treatment of depression. From 1 to 4 percent of the patients (over 1,800 in the study) reported agitation, diarrhea, dry mouth, delayed ejaculation, headaches, nausea, somnolence, and tremors. There have been at least two cases of men without a history of psychiatric disorders developing severe behavioral symptoms upon termination of paroxetine. The side effects were experienced not only while taking the medication but also after its termination and should cause some concern among clinicians.

SSRIs are also widely prescribed to treat symptoms other than depression. In fact, at least 25 percent of hospitalized cancer patients suffer from major depression, compared with approximately 6 percent of the general population. In 1992, a Canadian research group[44] studied the antiestrogen binding site that binds tamoxifen, a drug used in the treatment of breast cancer. In order to examine the site, the researchers used the compound DPPE (N,N-diethyl-s-(4-(phenyl-methyl) phenoxy) ethanamine), which is similar to tamoxifen. They found that DPPE stimulated tumor growth as a tumor promoter rather than as a direct mutagenic agent. In addition, DPPE was also structurally similar to two commonly used antidepressants, amitripyline and fluoxetine. They investigated the tumor-producing properties of the two antidepressants and found that both caused more rapid tumor appearance as well as greater tumor size.

It is important to note that the antidepressants are not direct carcinogens but rather may act as tumor promoters. This single study has caused considerable debate in the literature, and the general conclusion that seems to emerge is that there is a clear need for further epidemiological studies with humans to determine which antidepressants cause, promote, or inhibit cancers.[45]

It has been estimated that half of all individuals who commit suicide suffer from depression. Of patients diagnosed with a major depression, one-fourth will attempt suicide during their lifetime, and depression is implicated directly in approximately 15 percent of all suicides. One of the real dilemmas faced by physicians is that, of all the drugs used in suicide by poisoning, antidepressants are the most common. On one hand, antidepressants are very effective in treating depression, but they may also provide patients with a means to commit suicide.

Studies have shown that the frequency of suicides by poisoning with antidepressants varies widely with the particular medication being used. This finding does not address the cause of the differences between drugs. It is possible that it may simply be due to a higher rate of suicide attempts or a higher death rate per attempt. The results are complicated and somewhat confusing, but it appears there are some major differences in the toxicity of antidepressants.

Tricyclics have been found to have a higher death rate per overdose, due to the greater likelihood of cardiovascular complications, sedation, and suppression of respiration, when compared with nontricyclics, and in particular fluoxetine.[46] Tricyclic antidepressants accounted for 35.5 deaths per million prescriptions, while SSRIs accounted for 1.7 deaths per million.[47] In fact, the emerging consensus is that SSRIs, particularly fluoxetine, show no increased risk for suicide, but even the slight increase in risk associated with tricyclics is trivial when compared with the effectiveness of antidepressants in preventing suicide. In other words, given the effectiveness of antidepressants in ameliorating depression, there is no justification for not using tricyclics as well as SSRIs.

Prozac was the first of the SSRIs on the scene, but there are now at least seven more competing for their share of the market, particularly since in August 2001 the patent held by Eli Lilly for Prozac expired. Paxil™ (paroxetine), Zoloft™ (sertraline), Serzone™ (nefazodone), Celexa™ (citalopram), and Effexor™ (venlafaxine) have also received widespread acceptance as effective medications for treating depression. Interestingly, the *Physicians' Desk Reference* notes essentially the same warning about the abuse potential for all: none of the currently available SSRIs noted above have been systematically studied in animals or humans for their potential for abuse, tolerance, or physical dependence. While clinical trials did not reveal any tendency for drug-seeking behavior, these observations were not systematic, and it is not possible to predict, based on this limited experience, the extent to which a psychotropic drug will be misused, diverted, and/or abused once marketed.[48] The exact same warning was issued for each of the other SSRIs. This simply means that the long-term effects of these drugs are unknown. Safety is inferred from the lack of reports of drug-seeking behavior during clinical trials, not from systematic experiments. While this should be a cause of concern, it is important to emphasize that the benefits derived from these drugs may outweigh the potential for serious adverse consequences.

One federal agency that is concerned about the side effects of antidepressants, including the SSRIs, is the Federal Aviation Agency (FAA). The FAA will not certify any antidepressant medication, nor are pilots allowed to fly if they are taking antidepressants. The FAA requires that a pilot must be off medications and remain without significant depressive symptoms for several weeks to several months, depending on the individual case. A pilot must have a psychiatric evaluation and a clinical summary, and periodic progress notes must be sent to the FAA by the attending physician if a pilot is treated for depression.

COSTS OF DEPRESSION

In addition to the drastic reduction in the quality of life for those afflicted with untreated depression, as well as for their families, depression leads to substantial mortality and morbidity. Depression imposes a significant economic burden on society due largely to its high prevalence, underdiagnosis, and undertreatment. A 1996 study estimated the annual cost of depression in the United States at approximately $43 billion. Direct costs of the disease are estimated at $12 billion, while indirect costs are estimated at $31 billion. Indirect costs are composed of $23 billion due to absenteeism and lost productivity in the workplace, combined with $8 billion due to premature death. This makes depression one of the ten most costly diseases in the United States. For purposes of comparison, AIDS has been estimated to cost $66 billion per year and coronary heart disease $43 billion. Depression costs more than chronic lung disease, estimated at $18 billion

per year. The economic burden of disease in the United States is staggering, but, through proper diagnosis and treatment, costs associated with depression could be considerably reduced.[49]

Along with the potential side effects and lack of information concerning
the long-term effects of SSRIs on drug-seeking behavior, there is also some
concern about the cost-effectiveness of SSRIs when compared with the tricyclic and other antidepressants. SSRIs are costly, in some cases over one
hundred times more expensive than tricyclic antidepressants for a twenty-
eight-day treatment period.[50] For example, the least expensive tricyclic antidepressant, imipramine in its generic form, ranged in price from $2 to $12,
depending on the dose, for a one-month supply. On the other hand, the
most expensive SSRI, fluoxetine (Prozac™), ranged in price from $71 to
$280 across the range of therapeutically effective doses.[51] When these dif-
ferences are multiplied by the number of people taking these medications,
the economics of antidepressant pharmacology are staggering.

Of course, the mitigating factors in the choice of SSRIs over tricyclics are
the reported better patient compliance and better recovery rates with
SSRIs. Compliance with a drug treatment regime is difficult to measure at
best. Compliance rates for all drugs combined have been estimated at 10 to
90 percent.[52]

A study of psychiatric outpatients in Great Britain found that 46 percent
of the patients on tricyclics were noncompliant after four weeks of treatment.[53] Comparable data for SSRIs show rates of compliance approximately equal to tricyclics.[54] A high rate of compliance is an important factor
in the decision to use a particular drug, but it is not the only factor. Drug
cost should be taken into consideration as well. Imagine that one drug costs
$3 per treatment but is associated with fifty fatal poisonings per million pre-
scriptions. On the other hand, imagine a drug that is associated with no fatalities but costs $33 per treatment. Avoiding the fifty fatalities in the one
million prescriptions would cost $30 million in additional treatment costs,
or $600,000 per life saved. This may seem like a crass way to look at the
cost-effectiveness of treatment, but when data on the cost-effectiveness of
particular antidepressants are scant at best, it is not unreasonable to consider drug costs.

ARE WE REALLY THAT DEPRESSED?

Much of this chapter has focused on the widespread nature of depression and
the suggestion, nevertheless, that rates of depression, particularly in Western
society, are grossly underdiagnosed; that is, most people who actually suffer

from depression are not recognized as being depressed by their physician. Alternatively, there is the seldom-considered possibility that patients with minor mood disorders are unnecessarily treated for symptoms that would have disappeared spontaneously had intervention not been initiated.[55]

A substantial proportion of the prescriptions for antidepressants in the United States are written by physicians in general practice, pediatrics, or internal medicine, not by psychiatrists. Today primary care physicians are caught in the middle. Increasing economic pressures force physicians to see more and more patients, with the net result that each patient is seen for increasingly shorter periods.[56] When confronted with a patient who is clearly in pain but is unwilling or unable to see a mental health professional at that time, the physician wants to do something to eliminate the suffering immediately. When the diagnosis is unclear, the physician may very well prescribe one of the SSRIs because of their overall safety. While this is the practical reality in day-to-day life for a number of primary care physicians, it is the symptom of a much larger problem for the health care industry in particular and for society in general. We are unwilling to recognize the reality of mental illness, provide adequate services for the treatment of mental illness, provide the funds to pay for treatment, and invest in research into the causes of many mental illnesses.

There are some interesting similarities in prescribing patterns between physicians in the United States and the United Kingdom, particularly in regard to antidepressants. Over 95 percent of patients treated for depression in the United Kingdom are treated by general practitioners, not by psychiatrists, and the most frequently prescribed antidepressants are the tricyclics and SSRIs. With the costs of drugs rising 14 percent per year, there is considerable pressure to reduce expenditures.[57] Consequently, physicians tend to prescribe the lowest possible effective dose. One of the problems is that the prescription of antidepressants at less than therapeutically effective doses results in considerable waste. Particularly for the tricyclics, with over 85 percent being prescribed at low doses, patients were failing to enjoy relief from the symptoms of depression. The failure to treat depression early actually contributes to the development of recurrent and chronic depression. On the other hand, SSRIs were prescribed at effective doses in 98 to 99 percent of the cases. There are likely a number of contributing factors to these differences, but the conclusion from the analysis of prescribing practices is that SSRIs should be used as a first-line of treatment for depression.

Even though SSRIs have been shown not to be clinically more effective in treating depression than the older tricyclics, they are being prescribed at a rapidly increasing rate, primarily because of their more favorable side effect profile and their high degree of safety in the case of overdose. This widespread consumption of antidepressants is big business. The beneficiaries of these new designer drugs are the pharmaceutical companies that hold their patents. Given the size of the economic stakes in the prescription and sale of antidepressants, it should come as no surprise that pharmaceutical companies spend millions of dollars each year on marketing their products.

ADVERTISING AND DRUGS

A survey of the pharmaceutical industry in Great Britain found that about £300 million are spent annually on advertising. It has been estimated that pharmaceutical companies budget upward of £2,000 per physician for drug promotion.[58] In the United States, spending by pharmaceutical companies is increasing at an extraordinary rate. While drug advertising is permitted in many countries, it is typically restricted to medical publications and other print media. But in New Zealand and the United States, pharmaceutical companies are allowed to advertise their products on television. In the United States, drug advertising in all types of media is regulated by the Food and Drug Administration (FDA).

In 1981, the FDA established that "advertisements must present true statements relating to side effects, contraindications, and effectiveness" and that advertisements not meeting these requirements would be considered false and misleading.[59] In August 1997, the FDA issued a clarification of its policy about drug advertising. Until then, the FDA had required that manufacturers include nearly all of the consumer warning label in any advertisement, a requirement that made it impossible to use a thirty-second television spot. The 1997 clarification allows television commercials to name only the product and the disease, as long as viewers are given information about the "major" risks of the drug and are provided with information about where to go to get additional information. This policy shift has had a profound impact on the drug industry. Pharmaceutical companies have made record profits in the past few years, and this unbridled success seems to be related directly to their advertising expenditures.

Advertising has the intended effect of exposing consumers to particular drugs for common ailments. Patients can then see their doctors and ask for the brand-name drugs. Consumer spending on drugs went up almost 19 percent in 1999, reaching $131.9 billion, according to a recent market survey. The rise in spending is due in part to the changing demographics in the

United States, but, more important, the increase is also partially due to a rising volume of brand-name prescription drugs.

Drug companies spent nearly $2 billion on advertising in 2000, and much of this was directed toward the television audience. It appears that from the drug companies' perspective the advertising paid off handsomely. For example, consider that for the antidepressant Paxil™ (paroxetine hydrochloride), manufactured by Glaxo Smith Klein, expenditures on advertising increased approximately 24 percent between 1999 and 2000, from approximately $6.5 million per month to slightly over $8 million. This boost in advertising spending resulted in a 24.5 percent increase in drug sales for 2000 ($1.8 billion in sales), as well as a 17.2 percent increase in the number of prescriptions written.[60] Even by Alan Greenspan standards, this is big business.

Pharmaceutical advertising is one of the major sources of information about medications to practicing physicians. Drug manufacturers typically supply material directly to physicians as well as advertise in leading medical journals. It seems particularly important that advertisements should conform to the FDA guidelines, but in reality there is little incentive for publishers of professional medical journals to scrutinize the ads. In a study of the accuracy of scientific information presented in pharmaceutical advertisements, two physicians and a clinical pharmacist rated 109 advertisements in ten leading medical journals and found:

1. In 30 percent of the cases, the experts disagreed with the contention in the advertisement that the drug was "the drug of choice" for a particular condition.
2. In 32 percent of the ads, headlines misled readers about efficacy.
3. In 44 percent of the cases, experts felt that the ads would lead to improper prescribing if a physician had no other information about the drug except that found in the advertisement.

In sum, the study found noncompliance with FDA regulations in 92 percent of the advertisements. In the estimation of the experts, almost two-thirds of the ads needed major revisions. Now imagine, if the ads in professional medical journals had these problems, what kind of problems could be found in television advertisements?[61]

I am not suggesting that physicians are consciously making choices for prescriptions with little or no basis in fact. However, I am suggesting that physicians are bombarded with propaganda about all sorts of drugs. The advertisements by pharmaceutical companies are targeted specifically for their audience and use a variety of sophisticated marketing strategies.

These ads are effective in increasing the rate at which particular drugs are prescribed. Many of them may be misleading or at least do not conform to FDA guidelines. Taken together, these findings suggest that the popularity of some of the SSRIs may be due to the advertising practices of their manufacturers and not necessarily the action of the drug itself. While there are subtle differences in the action of SSRIs, they all have much in common. In fact, the choice of one over another might be considered arbitrary, since they all have roughly the same side effects. So, what leads a physician to choose one of these medications over another? Powerful imagery in advertising is directly targeted at the physician—the potential prescriber—as well as directly at the patient, suggesting that a particular drug will give the prescriber more control over the patient's illness.

On the patient's side, there is also cause for concern. Take Paxil for instance, an SSRI that has been marketed to treat social anxiety disorder (SAD), which the ads describe as "an intense, persistent fear and avoidance of social situations." Like the clinically depressed, those who suffer from SAD are in need of pharmacological help, but that is not the same as a case of the jitters before making an office presentation. In this context, Paxil could be seen as a shyness pill. Eli Lilly has also gotten into the act, marketing Sarafem™ (fluoxetine hydrochloride) as a treatment for premenstrual dysphoric disorder (PDD). Remember that fluoxetine is our old friend Prozac, but now it is being marketed to treat two separate conditions. A television ad for Sarafem shows a typical supermarket scene with a frustrated woman pushing a shopping cart. It is possible that one could interpret the ad as hawking a pill that will overcome the frustrations of everyday life.[62]

Some may claim that I have been overly critical of the pharmaceutical industry, but I want to be clear. I do believe that antidepressants have played a dramatic role in reducing the suffering of a substantial number of clinically depressed patients. In particular, SSRIs have helped a large number of depressed patients who otherwise would not ever have been diagnosed, much less treated. For these people, antidepressants can be seen as life-giving drugs.

There are a number of patients, however, particularly in primary care, who do not ever see a psychiatrist but nonetheless rely on psychotropic drugs. It is this population that is "at risk." This is not to say that they are in serious danger from taking antidepressants. Clearly, they are not, since even extremely high doses of the SSRIs are relatively safe, although little data exist on the long-term effects of treatment with SSRIs. What I am concerned about are those people who are not truly clinically depressed but who just want to feel better, thanks to so-called designer psychopharmacology. Of course, there is nothing wrong with wanting to feel good. Should these patients be denied the medication that has been so successful in altering the mood of the severely depressed? Maybe not, but the danger is in reinforcing the feeling among the public that there are medications that can fix whatever ails you.

Antidepressants are only one of a number of medications touted as quick fixes for all sorts of perceived maladies. Depression, anxiety, nervousness, and irritability are all a part of everyday life, not some disease that must be cured by pharmacological intervention. The danger in this "take a pill and fix it" mentality is that people will simply treat the symptom and not ever address the problem.

DEPRESSION IS OK!

As I have argued in this chapter, depression is a highly evolved psychological predisposition. Under a particular set of environmental contingencies, it makes perfect sense to be depressed. Being depressed is a powerful emotion that should signal to us that something is wrong in our lives and we need to fix it. From an evolutionary perspective, being depressed is a psychological mechanism that helps us deal with defeat and loss. Depression should be a powerful signal to us to change our behavior, but what if we are able to cure the depression without altering our behavior? This is where there should be a delicate interplay between traditional psychotherapy and pharmacological intervention.

It has been demonstrated that the most powerful clinical technique for dealing with depression is a combination of approaches. When "talking therapy" is combined with drug treatment, the chance of severe recurrence is significantly reduced. Why is combination therapy not the procedure of choice? It seems that the answer to this question goes back to the unwillingness of our society to deal with mental illness in the same way that we deal with other types of clinical conditions. Psychological maladies have long been regarded as simply weaknesses of character. People suffering from depression, drug abuse, alcoholism, and so on were suffering simply from a lack of character. If they really wanted to, this line of thinking tells us, they could stop being depressed, drinking too much, or smoking. This attitude reflects a deeply hidden assumption about the human condition, which is that we are lacking willpower. This is simply not true. Ask anyone who is clinically depressed: Would you change your psychological state if possible? I daresay that few would reply they like being depressed.

Depression is a major disorder that requires considerable effort on the part of the patient and help from trained clinicians. Historically, however, Americans have been unwilling to pay the price to help those with mental disorders. No one thinks for a minute about the possibility of denying care to individuals afflicted with lung cancer, even if they smoked, or to those with malignant melanoma, even if they spent their lives sunbathing. In order to deal with this epidemic of depression, we need to understand that our psychological state is the outcome not only of our basic biology but also the cultural environment in which we find ourselves. We must to be willing to treat mental disorders the same way that other clinical conditions are treated, remembering that they are yet another product of our biology and our culture. Otherwise, much of the clinical medicine that is practiced today is nothing more than treating symptoms but not underlying causes.

One of the primary discriminatory agents in this conspiracy against mental illness in general, and depression in particular, is the insurance industry. While willing to pay for medication to treat depression, many insurance companies are unwilling to treat "talking therapy" the same way as other types of therapy. Why? This is a complicated question, and, as one would expect, there is a complicated set of answers. One possibility is that unwillingness to pay for certain types of treatment may be nothing more than an economic decision. In reality, insurance companies have the same goal as other businesses in our capitalist society, to make a profit. So then, treatment of illness becomes a business decision. Profits are a powerful motivating force, and insurance companies cannot be faulted for attempting to maximize them. What is troublesome, however, are the practices of some companies that would lead one to believe that they actually care about the insured's well-being. If, as in the case of "talking therapy," the prognosis is unclear and is revealed only with treatment, insurance companies will not want to risk profits on what they consider a marginally effective therapy. Driven by profit motives, insurance companies orchestrate the American health care system, determining preferred courses of treatment, length of hospitalization, and the relative value of certain diseases and certain treatments.

There is no quick fix for the treatment of depression. Psychopharmacological treatment is quite effective in alleviating the symptoms of depression in a substantial number of patients, but clinical studies have shown that it must be continued for a considerable period of time to be truly effective. In addition, the best patient outcomes are achieved by combining drug treatment with traditional psychotherapy. Depression can be a chronic condition. Relapse is increasingly likely with advancing age and severity of depressive events. Yet insurance companies have a willingness to pay for pharmacological treatment, treating the symptoms of depression, but not psychotherapy, which attempts to get at its causes. Such a policy has served only to fuel the fires of enthusiasm for medications that will make us feel better but will not necessarily cure us.

CONCLUSIONS

A number of factors have conspired to create the epidemic of depression in the United States:

- the relative effectiveness and safety of the SSRIs;
- the lack of major short-term side effects of SSRIs;
- the general dysphoria among many people of all ages;
- the tendency to want an immediate fix for any malady;
- increased economic pressure on physicians to see more and more patients in less and less time;
- highly sophisticated pharmaceutical advertising, not only in medical journals but in a variety of popular publications as well as on television;
- a lack of public understanding about mental disorders and the associated prejudice;
- the avarice and greed of many insurance companies.

Rather than being willing to experience sadness and even short-term depression as a normal and highly adapted psychological mechanism, we simply want to feel better immediately. In reality, it took over twenty-five years of widespread use before we understood the true nature of the benzodiazepines and their effects on human behavior. Who knows how long it will take us to generate a real understanding of the psychopharmacology of these new antidepressants, and at what costs?

6 | Welfare, Cooperation, and Evolution

> As the reasoning powers and foresight of the members became improved, each man would soon learn from experience that if he aided his fellow-men, he would commonly receive aid in return. From this low motive he might acquire the habit of aiding his fellows; and the habit of performing benevolent actions certainly strengthens the feeling of sympathy, which gives the first impulse toward benevolent actions.
>
> —Charles Darwin, writing about the evolution of "Moral Faculties" in *The Descent of Man*, 1871

The mismatch between what is and what ought to be is a problem that has confounded philosophers for a long time. However, any discussion of modern human behavior and evolutionary theory must be prefaced with the disclaimer that modern evolutionary theory does not take a position on what ought to be. Instead, the task at hand is to understand what is. In this chapter, I make a case for understanding what is and how the situation might be changed in a way that is consistent with evolutionary theory.

Much of what we see in nature is, upon close examination, quite repulsive. Among the Diptera (true flies), *Cochliomyia hominivorax*, commonly known as the screwworm fly, has evolved a particular life history that seems ghoulish to us, but it serves them quite well. In fact, it is precisely because of their evolved life cycle that in 1932 there was a serious infestation of screwworm flies in the southern United States, and by 1934 over 200,000 animal deaths, including approximately one hundred human victims, had been reported. The death of livestock around the world attributable to screwworm flies has significant economic and political effects. Today, screwworm flies have largely been controlled through the release of sterile females into wild populations.

The species name, *hominivorax*, means "man-eater," and was given by French physician Dr. Charles Coquerel in 1858 to flies that caused the deaths of hundreds of prisoners on the Devil's Island penal colony, French Guiana. The adult screwworm fly is about ten millimeters (0.4 inches) long, has a reddish face and three black stripes on its thorax, and ranges in color from dark metallic blue to metallic bluish green to metallic green. Females, once they are fertilized, lay their eggs, up to 250 at a time, on living animals on the skin near open wounds. Once a wound has become infested with larvae, it also becomes more attractive to female screwworm flies ready to lay eggs. As a result, a wound may become infested with hundreds to thousands of larvae. The egg masses look like fish scales, and larvae hatch from the eggs within a day. The larvae crawl into the wound and begin tearing at the host's tissue with a pair of sharp mouth hooks.

The maggots usually feed on wounds first, then quickly migrate to healthy tissue. The maggots have toxic saliva that promotes the infection of feeding sites. The flesh at the infected site is therefore more digestible to the maggot. The infection also causes the production of foul-smelling pus. This odor attracts other females that are about to lay their eggs to the infected site.[1] Mature larvae are tapered and about seventeen millimeters (0.67 inches) long and mature in four to ten days after hatching. Then they are shed from the host. The larvae pupate in the soil for three to fourteen days and emerge as adults. The entire life cycle takes about three weeks, and flies can produce eight to ten generations over a summer.[2]

In densely populated areas in many parts of the world, screwworm flies have been reported to lay eggs not only in human wounds but also in the rectum and vagina as well as in the nose of individuals with respiratory infection. In isolated cases, larvae have been reported to eat through the bone lining the inside of the nasal cavity and lodge themselves in the brain. This life cycle is one that most would find repulsive, and it might be tempting to place a value judgment on the hapless screwworm fly that would be totally inappropriate. Evolution and natural selection have operated on Diptera for millions of generations and have selected individuals with a particular combination of characteristics that has allowed them to survive, no matter how repugnant the behavior of the fly may seem to us.

It is of no consequence to the forces of evolution that some of us may feel a bit squeamish about the screwworm fly's life history. The point is that evolution does not necessarily produce what is beautiful, sweet-smelling, or nice. Evolution solves a particular problem with variation in existing elements, no matter how the solution offends our sensibilities. In the discussion of human behavior that follows, it is important to keep in mind that natural selection and evolution are simply solving a problem, not taking a moral stand.

THOUGHTS ON A NATURAL PHILOSOPHY OF HUMANS

One of the paradoxical characteristics of modern humans is their ability to engage in self-sacrificing altruism. That is, in strict terms of evolutionary anthropology, humans frequently enhance the ability of their reproductive competitors to pass their genes on to future generations. Some believe that this is the essence of being human. On the other hand, philosophers, social scientists, theologians, as well as scholars in general have commented on the essential conflict for humans, that is, overcoming the "natural" tendency toward selfishness. In this context, "natural selfishness" means that organisms, whether they are pine trees, polar bears, penguins, or *Planaria*, all have a predisposition to behave in their own best interest at the expense of reproductive competitors.

This debate over the essential nature of humanity has deep roots in a variety of intellectual traditions. As with almost any issue of intellectual importance, we can trace a concern about the fundamental nature of humans back to Aristotle, who felt that humans were by nature social and cooperative creatures and were separated from all other animals by a penchant for following "reasonable" patterns of behavior.

The dichotomy between the selfish and the cooperative nature of humans is best exemplified historically in the controversy between John Locke and Thomas Hobbes. Hobbes wrote that humans were unlikely to act using reason, because "of their perverse desire for personal profit." On the other hand, Locke said that humans had the potential to exhibit action guided by reason if one were able "to deny himself his own desires, cross his own inclinations, and purely follow what reason directs as best, tho' the appetite lean the other way."[3]

The focus of the debate in the eighteenth century was over the fundamental rationality of mankind. This idea of rationality is an interesting one, for it permeates the social sciences today. Rational Choice Theory (RCT) is certainly de rigueur, not only in departments of economics but in political science and sociology as well. The central theme of this intellectual tradition is that people assess the costs and benefits of some action and base their own actions on the outcome of that analysis. If the benefits outweigh the costs, so the theory goes, people are likely to engage in a particular behavior—make that decision, buy that product—make a rational choice. RCT has its origins in the writing of Adam Smith (e.g., *Wealth of Nations*).[4]

According to Robert Frank, a professor of economics at Cornell University, there are several variations on RCT. He discusses them in his book *Passions Within Reason*.[5] Present Aim Theory (PAY) suggests that people behave in a way that is the most efficient pursuit of whatever goals one has at the moment. The problem with this variant is that it explains virtually any behavior as rational simply because the actor prefers it. Committing suicide could be rational simply because the person prefers to do it. The other variant on the theme, Self-Interest Theory (SIT), asserts that an act is rational if it promotes the interests of the actor.

Imagine that you are traveling away from home and you have dinner at a fancy restaurant. Upon completion of your meal, you are presented with a bill that you gladly pay; however, you pause when you are calculating the tip. It is unlikely that you will be back in this town in the foreseeable future, and it is also unlikely that you will have dinner in this restaurant again. Even if you do, what are the chances that you will be served again by the same waiter? And, even if you were, what are the chances that the server would remember you and treat you according to the size of the tip you left? You may be tempted not to leave a tip at all, or at best only a dollar, but you decide to follow the typical convention today and leave a tip equivalent to 15 percent of the bill. SIT would argue that you did not behave in your own self-interest by leaving a tip at the restaurant. Even though you are likely to feel guilty if you stiff the waiter, SIT would argue that the rational thing would be to not leave a tip and live with the guilt.

To these two variations on the theme, Frank added the Commitment Model (CM), which relies on the observation that people do not always do what is rational, because behavior is orchestrated largely by emotions.[6] Emotions are not rational, but, as Frank argues, they have evolved. Frank says that emotions have evolved because, unlike rational self-interest, emotions favor long-term self-interest, independent of the short-run outcome. Emotions to Frank are "moral sentiments," essentially problem-solving devices designed to make us utilize our social relations in service of our long-term interests.

The idea that we bring some additional baggage to the cost-benefit analysis of particular courses of action is a good place to begin to talk about the apparent irrationality of some human behavior. Using Frank's language of irrational commitment, we can see numerous examples where humans forgo short-term advantage. This apparently altruistic behavior can be explained by the role of emotions in altering the payoff matrix of the interaction. Emotions like guilt help keep cheaters in check, even though the probability of detection is low; love triumphs over lust in maintaining long-term relationships;[7] jealousy holds infidelity in check; shame tarnishes reputation; and empathy elicits empathy.

What Frank is saying is that emotions have evolved because of their fitness-enhancing properties. If one acts in an altruistic and cooperative way, one will garner rewards from society. Forgo short-term gain, and you will be more than compensated in the long run for your temporary loss. In other words, short-term selfishness may pay more than short-term altruism, but long-term altruism pays much more than short-term selfishness. A true "nonemotional" assessment of the costs and benefits of an altruistic act would weigh in favor of short-term selfishness every time. With the rise of emotions, the calculus of the cost-benefit analysis changes considerably, with short-term selfishness outweighed by long-term altruism.

The general form of the argument suggests that individuals who lack emotion behave in a coldly rational manner, weighing the immediate costs and benefits without regard for the long term. In some ways, this is exactly the way that natural selection works, favoring immediate solutions to problems with little regard to long-term costs. The paradox, of course, is that humans do not follow this algorithm all that well. We are likely to engage in behavior that places us in mortal danger in order to save a nonrelative, share much-needed food with an unrelated stranger, and even forgo reproduction altogether to devote all of our energies to helping those in need. This behavioral dilemma is an essential part of our humanity today.

On one hand is the side of human nature that is selfish, acting in accordance with the rational choice model. And on the other hand is the side that acts in tandem with more long-term considerations. Human decision making, according to some, is not based on the coldly analytical model but has been redesigned to use emotions as a long-range guidance mechanism to aid in resource allocation decisions for future contingencies. This distinction about the fundamental nature of humans has important implications in our daily lives and for our discussion of our biology and culture.

NATURALISTIC FALLACY

The model of the world that one employs has important consequences for life on a daily basis. If we look at a bit of cultural history, we can see that even within the twentieth century there are important junctions where the differences in what people thought ought to be, and what was, played a role in larger sociopolitical decisions. To put this a bit differently, are humans basically selfish and in need of strong outside influences and pressure to behave in cooperative ways, or are we essentially Rousseauian ingénues corrupted by society?

There have been critical periods in American history that have favored one of these perspectives at the expense of the other. World War I was a time when the United States first became involved in the realities of the modern world. United States participation with the Allied forces in the war effort signaled the end of the age of American isolation and the beginning of the United States as a world power. It is no accident that after World War

I the people in the United States were receptive to new ideas about a new identity.

The war had taken a terrible toll on the United States, not only in terms of casualties and economics but emotionally as well. The end of the war brought dramatic economic upheaval as industries that had been mobilized for the war effort shrank as quickly as they had expanded. People moved from job to job. In the years immediately after the war, the cost of living rose at a precipitous rate. For the first time in American history, the word "strike" had real meaning in the vocabulary. Labor unions had gained a stronghold during the war, and afterward they continued to push to maintain the rule of "closed shops." Management, on the other hand, denounced the idea and called for the "American plan," or open shops. Meanwhile, the Russian Revolution, the rise and expansion of communism, and a few isolated bombings in the United States gave rise to the Red scare and considerable antiforeign sentiment.

There were many aftereffects of the Red scare. It was clear that the lofty ideals Americans held at the beginning of the war were naïve at best. The passion for war had been fueled from many sources, and in some corners it continued to burn hot even after the Treaty of Versailles. Instead of simply laying blame on the Axis powers, a new enemy was found. The Sedition Act was an attempt to repress anyone resisting the war or spreading doubt about the moral imperative that drove it. Consequently, the political left in the United States suffered a major defeat in the years after World War I. Most famous perhaps was the imprisonment of Eugene Debs, the leader of the Socialist party, under Wilson's Sedition Act. Intense nationalism and distrust of individuals of "foreign stock" were rampant.

In 1919, the nation adopted the Eighteenth Amendment to the Constitution, prohibiting the manufacture, sale, transportation, import, and export of alcoholic beverages. Prohibition became a divisive issue in the United States, with both sides of the debate tying alcohol consumption to larger social issues. Antipathy toward the growth of cities (the presumed scene of most drinking), the evangelical Protestant middle class, anti-alien and anti–Roman Catholic sentiment, and rural domination of the state legislatures all conspired to bring about passage of Prohibition legislation. Prohibitionists tied alcohol consumption to the Roman Catholics, to immigrants, and to the quest for worldly pleasures. The debauchery of alcohol consumption was seen as a powerful threat to the morals of the nation. Other arguments for Prohibition included the corruption existing in the saloons and the industrial employers' growing concern for preventing accidents and increasing the efficiency of workers.

Those who sought to repeal Prohibition, on the other hand, associated Prohibition with Protestant discrimination, racism, and bigotry. For thirteen years, Prohibition was the law of the land. Illegal liquor manufacture and sale, the development of the speakeasy, and feelings of increased restriction on individual freedom ultimately brought an end to Prohibition, but not without leaving lasting scars on the American political scene. In 1932, the Democratic party adopted a platform calling for repeal of Prohibition, and Roosevelt's victory in the presidential election sounded the death knell of the Eighteenth Amendment.

The 1920s saw profound changes in American society. In literature, music, popular culture, as well as in the performing arts, the nation was undergoing a radical intellectual and emotional shift. Politics reflected the complicated nature of the social system at this time. Foreign relations were incredibly complicated, with treaties and acts being signed to give the illusion of stability at home. Casting a pall on the sociopolitical scene was the instability of the stock market. The stock market crash of 1929 and the ensuing economic decline saw the American economy reach its nadir by the beginning of 1933.

This side trip into American history sets the stage for what was to be a remarkable event in Western intellectual development. Margaret Mead was born in 1901, the eldest daughter of an academic family. She received her undergraduate degree from Columbia University in 1923. In 1925, she completed her dissertation and was awarded a fellowship to conduct research in American Samoa. Mead's ethnographic research on gender roles, adolescence, and child-rearing practices from a cross-cultural perspective would change the way the general public viewed anthropology.

Mead went to Samoa with an agenda, and, while there is certainly controversy over the accuracy of her data, that is a side issue at this point in the story.[8] Whether right or wrong, Mead came back to the United States after her fieldwork and then traveled in Europe, where she wrote *Coming of Age in Samoa*.[9] It became a best-seller, and for the first time exposed the public to the world of anthropology.

The central message that Mead was trying to get across was that human behavior was plastic and could be molded by culture. The attitudes toward adolescent sexuality in Samoa were so radically different from those in the United States that the only possible explanation was the role of culture in orchestrating human behavior. Mead emphasized the enormous variation in the solutions to problems that confront all humans, and the power of culture in directing local solutions to fundamental problems. More recently,

Derek Freeman has made a concerted effort to discredit Mead's work.[10] The net result of his efforts is the reluctant admission by some cultural anthropologists that the human mind is not a blank slate after all; there are some biological etchings on the slate that make certain messages easier to read and write than others.

Interestingly, Darwin recognized this debate about the essential nature of humans. He observed that, among some upper-class families, there were individuals who had a propensity to lie and steal. He concluded that since there was little economic motivation by the upper class to steal, the trait must be inherited. Moreover, he wrote, "If bad tendencies are transmitted, it is probable that good ones are likewise transmitted."[11] So in his own way Darwin was arguing for a biological basis for ethical behavior.

Ernst Mayr, a distinguished evolutionary biologist at Harvard, wrote that taking up either extreme position was a problem.[12] Mayr noted that there was considerable evidence supporting the cultural basis of cooperative or ethical behavior, including: (1) ethnic differences in moral or ethical behavior; (2) ruthless and amoral behavior toward slaves; (3) callous behavior during war, when civilian population centers have been bombed in the name of unavoidable collateral damage; and (4) the ease with which cultures can be radically changed in a single generation.[13] Mayr modified the extreme deterministic position of biology and combined it with recognition of the importance of cultural learning. He suggested that the most profitable way to think about ethics or morals is to think in terms of predisposition or capacity for adopting particular kinds of behaviors.

The title of this section promised a discussion of the "naturalistic fallacy," and I have gone the long way round to introduce the idea. The naturalistic fallacy is a deductive fallacy that tries to draw a conclusion about how things ought to be, based solely on information about how things are right now.[14] The naturalistic fallacy was proposed by G. E. Moore, a pioneering literary philosopher, in his book *Principia Ethica*,[15] published in 1922. He reasoned that the notion of moral goodness could not be defined or identified with any property. He attempted to show that statements like "this is good" are sui generis and cannot be reduced to statements of either natural or metaphysical fact; the idealist belief that ethics ultimately depend on metaphysics rested on a delusion.

We intuitively recognize goodness when we see it, but "goodness" itself cannot be defined. For Moore, philosophers who attempt to define intrinsic goodness commit the naturalistic fallacy, the fallacy of defining "goodness" in terms of some natural property, such as pleasure. What is pleasurable is good and what exists now that is pleasurable is what ought to be, is the short version of the story. This implies that all moral theories based on anything other than immediate moral intuition must fail. Whether an act produces pleasure, or is in accord with the will of God, or is conducive to reason is beside the point. The naturalistic fallacy serves as a justification for what exists.

The *converse* of the naturalistic fallacy justifies changing what exists for what ought to be, or what some might call political correctness. The idea that is central to the "antinaturalistic fallacy" is the changeability of behavior. So, if you find a behavior that people find pleasurable and engage in frequently, it is not necessarily what ought to be, and we have the power to change behavior accordingly. The "antinaturalistic fallacy" tells us that racism, chauvinism, elitism all exist and are within our power to change, if we can just be persuaded to do so.

This idea that human behavior is infinitely malleable goes back to the generation of scholars associated with Franz Boas, Ruth Benedict, and Margaret Mead. This malleability to change in whatever direction that local culture dictates was, and still is, a fashionable idea. We can overcome the lack of humanity in the world by simply persuading people to change. It is an unpopular reality that these kinds of changes may be beyond the evolutionary potential of the organism, and hence impossible, but more about that in a bit. This is not to say that we ought to ignore basic principles of human dignity and respect for others, but it is clear that what ought to be, in the minds of some people, is not what is, and likely may never be, in spite of vigorous protests to the contrary.

Two huge political experiments that have failed offer some insight into the plastic nature of human behavior. The fall of the Soviet Union and the manifest failure of the People's Republic of China to create the new "Leninized" man should give some pause to those interested in changing human behavior. B. F. Skinner, the American experimental psychologist who gained recognition for his invention of the air crib and the Skinner box,[16] and his disciples were advocates of a school of thought in psychology called behaviorism.[17] The behaviorists contended that, in its most radical form, the mind works as a black box. That black box enables the individual to learn any task with equal facility. By constructing a system of rewards and punishments, they said that human behavior could be molded in an infinite number of ways. For Skinner, in *Walden Two*, operant conditioning was the key to creating a utopian society.[18]

PRIMER OF EVOLUTIONARY THEORY

Before offering a few thoughts on how our evolutionary history might affect our altruism and modern systems of redistribution of wealth, consider what cooperative behavior might look like in animals and what sorts of explanations might be offered to account for it. Remember, what we are looking for are cases where animals appear to engage in behaviors that require an investment of time, energy, or other resources in other individuals rather than in themselves. These kinds of altruistic behaviors have long been known among those who study animal behavior.

Darwin, writing in *On the Origin of Species*,[19] recognized the possibility that animals could act in ways that would benefit others. He was fascinated with the sterile castes of certain social insects and marveled at why social insects could develop neuter castes, leaving no descendants to benefit from their efforts. Slave-making ants proved particularly troublesome for Darwin, and, upon initial consideration, he commented that they presented "one special difficulty, which at first appeared to me insuperable, and actually fatal to my whole theory."[20] The central question for Darwin was how a system that seemingly exploited one species could have ever evolved, and how such a system could persist in nature. A strict selectionist view would predict that the exploited species would become extinct. Similarly, how could species survive that produced individuals that did not reproduce?

Darwin managed to get around this central problem when he introduced the idea that natural selection could operate at the level of the family and not only on individuals. He reasoned that families that could produce sterile individuals that were helpful to the reproductive family members would be more successful than families in which reproductives were in competition with all others. While this helped Darwin out of the intellectual quandary in which he had found himself, the question has proven to be more of a problem than Darwin first envisioned. There is the puzzling fact, for example, that altruistic behavior sometimes reaches beyond families, benefiting non-kin neighbors or even strangers. President George W. Bush summed up this dilemma nicely in a campaign speech at an oyster roast in South Carolina when he said, "We must all hear the universal call to like your neighbor just like you like to be liked yourself."[21]

Evolutionary Theory: Nepotism

The existence of self-sacrificing altruism to non-kin was a problem largely ignored until a young evolutionary biologist at Harvard, Robert Trivers, wrote an important paper in the *Quarterly Review of Biology* in which he addressed this problem.[22] Until Trivers, evolutionary biologists recognized that self-sacrificing altruism could be explained via kin selection.[23] That is, individuals were predisposed to act in their own self-interest and could actually do so not only by producing offspring themselves but also by furthering the reproductive careers of a relative. The extent to which one should aid a relative ought to be dictated by the closeness of the relationship. Intuitively, one would be inclined to give aid to full siblings more quickly and at greater personal cost than to first cousins or geriatric, postreproductive aunts, for example.

While a graduate student at the London School of Economics, W. D. Hamilton, in a two-part paper published in the *Journal of Theoretical Biology* in 1964, demonstrated mathematically how such an altruistic trait could evolve in the population, and he called the process evolution by kin selection. The implication of this idea is profound, for it explains the often observed altruism among kin. Individuals can be favored in the evolutionary struggle if they produce offspring that, in turn, survive and reproduce themselves. But individuals can also gain genetic representation in future generations through the successful reproductive efforts of individuals with whom they share genetic material. From this perspective, helping a sibling, aunt, uncle, niece, or nephew makes perfect Darwinian sense.

Moreover, individuals would be expected to gain more from helping those to whom they are the most closely related. Similarly, one would expect that individuals would be willing to incur the highest costs when there was a potential to derive the highest payoff. While we share, on average, one-half of our genes with a full sibling, we share only one-quarter of our genes with an aunt or uncle and a mere one-eighth with a first cousin. At first glance such small fractions of shared genes may not seem like much of an incentive for self-sacrifice, but maybe I can explain why this payoff is such a big deal.

In order to explain what is going on here, I have to talk about primate evolution and molecular biology. Anatomists have long been concerned about establishing ancestor-descendant relationships among extinct species, and these relationships are usually diagrammed as a family tree. These family trees are called phylogenies and reflect the relationships between groups of species. In fact, there is a branch of biology called systematics that is devoted to developing precise rules for establishing phylogenies.

Until the 1970s, phylogenetic trees were based on anatomical evidence—the tale of the bones, so to speak. Molecular biologists and evolutionary anatomists had little to say to each other for decades, until the work of molecular biologist and ornithologist Charles Sibley became known. Sibley was interested in unraveling the evolutionary history of birds, which turns out to be very complicated because natural selection has solved the fundamental ecological problem of how to make a bird in many different ways. That is, if you look at birds and try to make comparisons based on feeding patterns, for example, you will find that birds that appear similar in feeding adaptations can have very different evolutionary histories. For example, American vultures and Old World vultures look a lot alike and make a living in similar ways, but American vultures are descended from storks and Old World vultures from hawks.[24] Anatomy was not very helpful in this case because it led to the erroneous conclusion that these two types of vultures had similar phylogenetic relationships. Sibley and his colleague, Jon Almquist, began to use a technique in molecular biology to try to unravel the phylogenetic history of birds.

Using a concept called genetic distance, molecular biologists can determine the genetic relationships between two organisms, based on the similarity of their DNA. Molecular data allow comparisons to be made on a wide array of characteristics, not just on a few characters represented in the hard-tissue anatomy. The fundamental assumption of this technique is that organisms that are genetically similar share a more recent common ancestor than those less genetically similar.

DNA hybridization is a technique used to establish these genetic distances. Scientists take tissue samples from two organisms of interest and extract the DNA. The DNA is cut into strands and mixed together, which allows the strands to bind together. The strength of the bond between the two strands of DNA is a direct measure of the closeness in their evolutionary history. If the bonds are strong, the two species share a similar genetic history. If the bonds are weak, then the relationships are more distant.

In order to quantify the strength of the bonds, researchers use heat to break them apart. As the DNA soup is heated, the weaker bonds will break at a lower temperature than the stronger ones. The temperature at which half of the bound DNA pairs break apart is a measure of how many identical DNA pairs there are between the two species. Researchers can then take samples from various species and compare them. In that way, they can estimate genetic relatedness. The genetic distances among humans and the great apes are shown in table 6-1.

In other words, these are the data that people are talking about when they say that we share 98+ percent of our genes with chimpanzees. This is an impressive statement that has led some to redefine our relationship with primates in general.[25] However, a couple of points are a bit deceiving in this

Table 6-1 Genetic Distances (%) between Humans and Ape Species

	Human	Chimpanzee	Pygmy chimp	Gorilla	Orangutan
Human	—	1.63	1.64	2.27	3.60
Chimpanzee		—	0.69	2.21	3.58
Pygmy chimp			—	2.37	3.56
Gorilla				—	4.28
Orangutan					—

Source: C. G. Sibley and J. E. Ahlquist, 1987, DNA hybridization evidence of hominoid phylogeny: results from an expanded data set, Journal of Molecular Evolution 26 (1–2), 99–121.

statement about our ancestry. First, when we discuss the similarity of species using this technique, the starting point for comparison is not zero percent but 25 percent. Since there are only four bases used in the construction of DNA, probability theory predicts that they would match by pure chance one out of every four pairs. Given that we share some fraction greater than 25 percent of our DNA with pine trees, for example, perhaps we should not make quite so much about the 2 percent difference between humans and chimpanzees.

On the other hand, the 2 percent difference is remarkable because that is where all the difference between chimpanzees and humans resides. That small bit of difference between human and chimp DNA codes for our bigger brains, bipedal posture, relative scarcity of body hair, spoken language, artistic and scientific achievements, and our sociopolitical system. All of these characteristics and more serve to separate us from other animals.

Now let us take this example a bit further. If we are separated from chimpanzees by less than 2 percent of our DNA, imagine the minute differences between populations of modern humans, say some fraction of 1 percent. Now imagine the differences in genetic makeup among your immediate family. When we talk about Hamilton's ideas of kin selection for altruistic behavior, we are inferring that there are mechanisms for extremely subtle discriminations among relatives encoded in our genome. Likely, some basic rules, rather than a specific instruction set, are encoded for each individual. For example, the instruction might be something as simple as "treat those with whom you were reared as close kin." The point is, if Hamilton is right—and most evolutionary biologists think that he is—we are capable of some amazing feats of discrimination of relatives from nonrelatives.

From the preceding discussion, it is easy to see how an individual could benefit in evolutionary fitness terms by helping a relative. In one sense, when you help a relative, you are helping yourself. Evolutionary biologist J.B.S. Haldane published a paper in 1955 in which he anticipated much of what Hamilton would say nearly a decade later.[26] In this paper, Haldane illustrated the importance of kin in the expression of altruistic behavior when he imagined that individuals possessed a gene that would predispose them to try to save a drowning individual, even if they might drown themselves. He noted that you would have to save a cousin eight times as often as you might drown in the attempt in order for natural selection to favor the gene. Another way of putting it is that you would have to save more than eight cousins in your self-sacrificial drowning for the gene to be favored, because we share with a cousin, on average, only one-eighth of our genes. The fundamental idea here is that of the 99+ percent of DNA shared in common among all humans, we share half of that tiny, tiny fraction with a full sibling, one-quarter with aunts and uncles and nieces and nephews, and one-eighth with first cousins. The amazing feat is the ability to make these kinds of subtle distinctions.

The importance of kin in shaping human social relations is profound. Historically, understanding kin networks has been the "meat and potatoes" of cultural anthropology.[27] Today, the collection of detailed genealogies is not necessarily a high priority in many ethnographic projects, but changes in intellectual fashion within cultural anthropology have not diminished the importance of understanding basic biological relatedness. However, there are a small but growing number of anthropologists who recognize the importance of biology in shaping the behavior of modern humans. These are anthropologists who are trained both in the biological as well as the social sciences and seek to understand the human condition as an interaction of biology and culture. It is not unreasonable to imagine that recognition of consanguinity has a long and distinguished history in our own evolutionary story. The power of kin recognition was likely an important organizing principal in the evolution of human social behavior.

Evolutionary theorists recognized the importance of kin relations in the study of animal behavior and have extended their analysis to humans. And why not? The strong predisposition to behave positively toward our relatives is often an important force in making important life history decisions. Certainly, our choice of mating partners is directed, in a nontrivial way, by our kin relations. Individuals who are eligible partners are defined by both our culture and biology. While arranged marriages are not a typical occurrence in Western society, relatives have a powerful influence on marriage partners. Going home to meet the parents is one of the first tests of a relationship. Parental disapproval may not preclude marriage or mating, but it does make a relationship more difficult in both the long and short run. The whole question of mate choice is a fascinating problem but one that is outside our present discussion.

Anthropologists have long recognized that eligible marital partners are dictated both by concern over political alliances and consolidation of wealth in some societies, and not necessarily by the emotions of the couple involved. A latent function of these kinds of marital proscriptions is a reduction in the deleterious consequences of inbreeding. Now the point of this discussion is not a discourse on human incest but a consideration of the importance of kinship in directing human behavior.

If Hamilton, Trivers, Haldane, and other evolutionary biologists are right, then kin relations are likely to be important in the reallocation of wealth. We would expect that humans would follow the well-known Hamiltonian idea of fitness maximization through dispersal of wealth first to progeny and then to other relatives, in proportion to their genetic relatedness to the benefactor. Patterns of inheritance have been studied in several different contexts, and the results are similar across a wide array of human cultures: kin are favored over non-kin, and close kin are favored over more distant kin.[28] Nepotism has long been a favored behavioral strategy among social organisms, humans certainly included. Given this strong predisposition to favor kin over non-kin, all things being equal, it should come as no surprise that we are not inclined to act altruistically toward those not related to us.

Evolutionary Theory: Mutualism

There are situations, however, in which individuals engage in altruistic behaviors with nonrelatives. The altruism in these cases is often referred to as "by-product altruism or mutualism," since both individuals involved in the interaction receive some benefit. Images of the coordinated hunting activity of chimpanzees come to mind as an example. Geza Teleki and more recently Craig Stanford, both trained as anthropologists, have published accounts of chimpanzee predatory behavior in the Gombe Reserve in Tanzania. Christophe and Hedwige Boesch have studied predatory behavior of chimps in the Tai National Forest (Côte d'Ivoire), and Japanese researchers have reported predation in the Mahale Mountains in Tanzania.

Details of the chimpanzee predatory behavior vary slightly from site to site, but the overarching pattern of cooperative hunting remains the same. Favorite prey in the Gombe, as well as other sites, is the red colobus monkey, a relatively small, highly arboreal, leaf and fruit eater of the middle and upper forest canopy in the forests of sub-Saharan Africa. The important point is the cooperative nature of the hunt. The calculus of the payoff to individual hunters is a matter of some contention among primatologists, but for our purposes this looks like selfish behavior as a result of "by-product mutualism."[29]

Other examples of mutualism include the side-by-side position of two horses in a pasture, while facing in opposite directions, so they can swat flies from around each other's head; foraging by red-billed oxpeckers on domestic cattle; cleaning of coral reef fish by wrasses; feeding associations between hippopotamuses, capybaras, and jacanas; nest associations in fish, plant, and insect pollination; and so on.[30] The thing that we are concerned with is the lack of relatedness between the interactants. In fact, one of the common characteristics of mutualism is that it frequently occurs between members of different species. I want to be clear that the focus of this chapter is not on this kind of mutualism but on what we recognize to be true "self-sacrificing" altruism.

ALTRUISM AMONG NON-KIN

The comparative perspective has contributed immensely to our understanding of human behavior. In fact, it is instructive to examine the self-sacrificing altruism in other species so we may more clearly understand what is going on in our own. What do vampire bats, olive baboons, sticklebacks, and coatis have in common? Okay, so I gave the answer away already. But are they really altruistic? Let us take a look at the data.

As I have suggested, altruism to a nonrelative makes absolutely no sense in the calculus of natural selection.[31] Nonetheless, all of us can think of examples in our own personal experience of what appears to be altruism among nonrelatives. For the true skeptics among us, one might argue that what appears to be altruism among unrelated individuals is actually some variant of kin selection with the actors only distantly related. Trivers recognized this possibility and set out three conditions that must be met for an interaction to qualify as reciprocal altruism:

1. Animals should have a reasonable chance of encountering one another in a social setting after the initial altruistic act.
2. Altruists must be able to recognize each other and thereby detect cheaters.

3. The cost (to donor)/benefit (to recipient) ratio must be low. Interactions in which the donors incur little cost and the recipients accrue significant benefits are more likely than the converse.[32]

One of the crucial aspects of the evolution of reciprocal altruism is that there must be a mechanism to avoid the "you scratch my back, and I'll ride yours" problem. In other words, there must be a way to detect cheaters. There is no way that reciprocal altruism can evolve if cheating is rampant. No one will be inclined to act altruistically if by doing so they suffer at the hands of the selfish cheaters. Now if all this sounds as if it requires organisms with great intelligence or at least big brains, let me dispel that idea.

Vampire bats (*Desmodus rotundus*) are very small mammals weighing only fifteen to fifty grams (0.5–1.7 ounces). They are native to an area extending from northern Mexico to central Chile, feed exclusively on fresh blood, and attack a wide variety of prey species. They typically go for areas where hair and feathers are scant or absent. When a suitable area is found, they make a quick shallow bite that is painless to the victim. Sleeping individuals are generally not even awakened. Vampire bats have a voracious appetite and may continue to feed for up to half an hour. An anticoagulant in their saliva may keep the wound open for as long as eight hours.[33] Like other bats, vampires are communal roosters, and, after a blood meal, upon returning to the roost, individuals will regurgitate blood into the mouths of roostmates. Now why is this a big deal?

In the world of mammals, vampire bats are wee tiny, and, when confronted with all the problems of small body size and the reliance on only one type of food, missing a meal can be deadly. Vampire bats must eat regularly, since a typical meal consists of only about twenty milliliters of blood. Seventy-two hours without food is fatal to a vampire bat. To make matters worse, vampire bats are not always successful when they go out to feed. In fact, about 7 percent of the adults and 33 percent of juveniles younger than two years of age fail to find food on a typical night. It is easy to see that sharing a blood meal, even with individuals who are not your kin, could have a big fitness payoff, particularly if the recipient one night might be the donor the next.[34] The whole system would break down if there were not mechanisms to recognize cheaters. We do not know what these mechanisms are specifically, but surely they must exist, because vampire bats all over the Americas continue to thrive.

Cases have been made for altruism among nonrelatives in several other species with varying degrees of success (e.g., white-nosed coati, *Nasua narica*, olive baboons, *Papio anubis*, guppies, *Poecilia reticulata*),[35] but, to date, the vampire bat example remains the most convincing. The public perception of cooperation among animals is at best naïve and at worst just plain wrong. In general, we tend to look at animals through the rose-colored Disney filter. The "feel good" filter tells us that animals work together for the common good, holding hands or paws and gathering in a circle to sing about the goodness of life. This naïve view of cooperation in animals is fostered by the "television nature program" mentality.

Take the portrayal of meerkats (*Suricata suricatta*) on television as an example. Meerkats are relatives of the mongoose and are about ten to fourteen inches long. They have pointed snouts, silvery-brown fur, and irregular dark stripes on the rump. Their faces are white with dark markings on the ears and around the eyes. Meerkats are found throughout sub-Saharan Africa, and, with just a little effort, they are relatively easy to film. They have excellent eyesight, and single individuals often stand guard over the entire colony, keeping a wary eye for aerial predators.

Initially, meerkats were portrayed as "noble" because one individual acted as a sentry while the others ate.[36] This apparent self-sacrificing altruism appealed to our idea of animals acting for the common good. However, recent research shows that sentries are not being altruistic at all; they are actually taking care of themselves. An individual will act as a sentry and stand guard only after having eaten. When a predator shows up, the sentry immediately runs and hides, making him (or her) least likely to be eaten.[37] While it is tempting to imagine elephants, panda bears, and puppies acting for the good of their littermates, or the group, or the species, this is simply not the case. Only under very special circumstances will animals act in a truly self-sacrificing manner, and, as we have seen, it has been convincingly demonstrated only in a single bat species.

HUMAN EXAMPLES

People are uncomfortable with explanations of behavior that focus on self-interest and, for a variety of reasons, want to see behavior in a more charitable light. This uncomfortableness is particularly true when it comes to human behavior. In general, most people think that altruism is good and selfishness is bad. We have seen how kin relatedness can play an important role in promoting altruistic behavior because, while someone else benefits from your self-sacrificing actions, they also share some of your genes.

Altruism is an ideal, while selfishness is to be avoided at all costs—these are themes that we have heard throughout our lives. Rabbis, pastors, priests, teachers, scoutmasters, parents, and grandparents have all extolled the virtues of selflessness. Much to their credit, we all know examples of self-sacrificial behavior in our own lives. It is important to note, however, that although we all have examples of this kind of behavior in our own experience, they generally are rare. The following are examples of this kind of behavior that have attracted considerable attention in the press.

Arland Williams

It is not exactly a household name, but some over age forty who are reading this may recognize the name Arland Williams. The following is an excerpt from the official National Transporation Safety Board (NTSB) report on the crash of an Air Florida jet:

> On 13 January 1982, Air Florida flight 90, a Boeing 737–222 (N62AF), was scheduled to fly to Fort Lauderdale, Florida, from Washington National Airport, Washington, D.C. There were seventy-four passengers, including three infants and five crew members, on board. The flight's departure was delayed about an hour forty-five minutes due to a moderate to heavy snowfall that necessitated the temporary closing of the airport.
>
> Following takeoff from runway thirty-six, which was made with snow and/or ice adhering to the aircraft, the plane at 1601 EST crashed into the barrier wall of the northbound span of the Fourteenth Street Bridge, which connects the District of Columbia with Arlington County, Virginia, and plunged into the ice-covered Potomac River. It came to rest on the west side of the bridge 0.75 nautical miles from the departure end of runway thirty-six. Four passengers and one crew member survived the crash. When the aircraft hit the bridge, it struck seven occupied vehicles and then tore away a section of the bridge barrier wall and bridge railing. Four persons in the vehicles were killed; four were injured.[38]

This certainly was a tragic accident, but what most people will remember are not the details of the crash but the image of Arland Wiliams, a forty-two-year-old federal bank examiner from Atlanta, Georgia. Williams was captured on videotape hanging onto the tail section of the downed jet, passing a safety rope suspended from a United States Park Service rescue helicopter to five other surviving passengers of the ill-fated jetliner. Williams repeatedly grabbed the rope from the helicopter and tied it around others floating near him. On the sixth return flight, the helicopter pilot could not find the extraordinary good Samaritan. Arland Williams had given his life in order to save five other individuals. It was several days after the disaster before the identity of the hero, who had been shown on national television, was known. Since he was the only victim of the crash to die by drowning, medical examiners concluded that Williams must have been the mystery man.

This is an example of a particular kind of behavior that, from an evolutionary viewpoint, is not easily understood. Why would Arland Williams give his life to save people to whom he was not related and, as a matter of fact, did not even know? We can directly rule out helping to increase genetic representation in the next generation. What possible explanation can be offered to explain this self-sacrificing kind of altruism?

Agnes Gonxha Bojaxhiu

Mother Teresa, whose given name was Agnes Gonxha Bojaxhiu, was born on 27 August 1910 in Skopje, Yugoslavia (Macedonia), the daughter of a businessman who owned a building company. She had one brother and one sister, and after her father's death was raised by her mother. In 1928, she joined a religious order and took the name Teresa. The order immediately sent her to India. A few years later, she began teaching in Calcutta, and in 1948 the Catholic Church granted her permission to leave her convent and work among the city's poor people. She became an Indian citizen that same year. In 1950, she founded a religious order in Calcutta called the Missionaries of Charity. The order, now with branches in fifty Indian cities and in thirty countries around the world, provides food for the needy and operates hospitals, schools, orphanages, youth centers, and shelters for lepers and the dying poor. For her work with the poor around the world, Mother Teresa received the 1979 Nobel Peace Prize and other awards including the Pope John XXIII Peace Prize (1971) and India's Jawaharlal Nehru Award for International Understanding (1972). Mother Teresa died on 5 September 1997.

Mother Teresa's altruism is legendary. She is known worldwide for her steadfast rejection of material goods and her insistence on the diversion of resources to those most in need. The Nobel prize committee noted, in its press release about Mother Teresa, "she had worked tirelessly to improve the plight of children and refugees." The committee placed special emphasis on the spirit that had inspired her activities as a tangible expression of Mother Teresa's personal attitude and human qualities. Conspicuous features of her work were respect for the individual and for his or her dignity and innate value. The loneliest, the most wretched, and the dying had, by her hands, received compassion without condescension, based on reverence for all humanity. Like Arland Williams, but in a much more expansive way, Mother Teresa performed astonishingly altruistic acts.

Jonas Salk

The son of Russian Jewish immigrants, Jonas Salk was born 28 October 1914 in New York City. His early work in microbiology and epidemiology set the stage for a lifetime of research. One of Salk's early sources of research funding was the National Foundation for Infantile Paralysis (NFIP), an important supporter of research into the causes and possible cure of poliomyelitis as well a supporter of victims of this disfiguring and often fatal disease. Contrary to what many thought at the time, it was a private foundation with no explicit government support. The NFIP gave rise in the later 1950s to the March of Dimes and from its inception relied on the media as a powerful tool for acquisition of resources in the war against polio.

Jonas Salk was interviewed by Edward R. Murrow on the CBS pioneering television news program *See It Now*. During the interview, Salk reported the discovery of a vaccine against polio. It is difficult for us to appreciate the magnitude of this research and the health implications of the vaccine.

Oddly enough, it was improved sanitary conditions that led to the spread of polio in the first place. Previously, people were exposed to small amounts of the virus as infants, when the risk of paralysis was reduced and there was subsequent opportunity to develop antibodies against the virus. As infant exposure decreased, however, and the disease began spreading in the older population, parents were admonished to keep their children well bathed, well rested, well fed, and away from crowds. Public swimming pools, movie theaters, camps, and schools were closed, drinking fountains abandoned, and draft inductions suspended. From 1950 to 1954, there were about forty thousand cases per year reported in the United States. This highly contagious disease left a path of devastation wherever there was an outbreak. The pressure for a "cure" was enormous, and the announcement of the Salk vaccine was greeted with unbridled enthusiasm. Now given all this hoopla about the discovery, Salk was certainly mindful of his vaccine's economic importance. But, when asked by Murrow in the CBS interview who controlled the vaccine, Salk replied, "Well, the people I would say. There is no patent. Could you patent the sun?"

In this time of rapid biotechnology advancements and the general climate of avarice and greed among many corporations, particularly most pharmaceutical conglomerates, it is hard to imagine that an individual would be quite so altruistic. Salk was likely more interested in improving the human condition on a worldwide basis than in immediate financial gain. Of course, the curmudgeon in me has to suggest that Salk might have used this incredible act of generosity to his own advantage in securing funds to support his later research as well as the establishment of his research institute in La Jolla, California. By increasing his prestige and influence in the nascent biomedical research community, he set himself up for future payoffs. But in all fairness, this does not seem to be the case. The development and use of the Salk polio vaccine appears to be one of those landmark examples of human altruism.[39]

RECIPROCAL ALTRUISM AND REDISTRIBUTION OF WEALTH

Redistribution of wealth has long been a part of human society. The tradition of *tzedaka* (charity) in Judaism has its origins in the Old Testament book Exodus and is a central part of Jewish life. The Gabbaim are men in the Jewish Orthodox congregation who are in charge of the funds used for charity. Those chosen to be Gabbaim are men with reputations beyond reproach. In fact, the Talmud offers specific guidance not only about giving but also about the proper attitude for giving.[40] There is a long tradition of charitable giving in all major religions, and it became institutionalized in cultures in both the West as well as the East long ago. A more secular version of altruism is an important part of modern American society and traces its origins back to the New Deal.

It is helpful to provide some historical context for our discussion of altruism and the distribution of wealth. For most people, redistribution of wealth brings to mind taxation and government-funded programs devoted to helping people who are unable to fully support themselves or earn a living. In any society, some people are unable to work, and it is generally recognized that the very young and the old, as well as those physically and mentally disabled, have limited capacity to do work. This recognition fostered the development of social support systems that are an important part of our evolutionary history and have contributed to our evolutionary success.

It is likely that welfare systems are formalized types of social support that have long been a part of humanity. We know from comparative studies of animal behavior that observations of aid or assistance to individuals who are injured are sparse indeed. It has been reported that in wild toque macaques, injured males received disproportionate grooming by juveniles.[41] Patricia Whitten, a biological anthropologist at Emory University, in her field study of female vervet monkeys in northern Kenya, reported that, after an infant broke its leg, its mother carried it long after the typical end of intense maternal care-giving behavior.[42] In captive animals, there have been reports of aiding invalid individuals by other group members in dwarf mongooses.[43] Captive chimpanzees, as well as rhesus monkeys, have been reported to pull individuals from experimental situations where there was risk from various aversive stimuli (e.g., electric shock, loud noise).[44] What is striking about the comparative data is the paucity of examples of aiding individuals in distress, which has a long history for humans.

In trying to understand the evolution of reciprocal altruism among humans, one must examine the fossil record. It is important to look to our past to see if there are early examples of aiding behavior. The first question that comes to mind is what such behavior would look like represented in fossilized bones.

In order to find such evidence, we need to think about bones in a way that may not be familiar to some readers. Bone is living tissue and is either spongy or hard. It consists of a chemical mixture of inorganic salts and various organic substances. Calcium phosphate and calcium carbonate are the principal constituents of bone that cause the hardness, while its elasticity is derived from gelatin, collagen, elastin, and fat. Although bone is among the hardest parts of the body (second only to the enamel on our teeth), it is sufficiently plastic that, if broken, it will regenerate and, if exposed to pressure, can be remodeled.[45]

Human paleopathologists are able to look at bones of individuals who experienced known injuries or diseases during their lifetime and determine the specific effects of trauma on the skeleton. For example, certain types of trauma will leave instantaneous marks on bones or teeth—intentional mutilation, complete or compound fractures, or weapon wounds. Other types of trauma that are the result of chronic stress or disease may be more difficult to detect.

One of the most intriguing observations made about our evolutionary past that sheds some light on the origins of altruism comes from a cave near the Great Zab River in northeastern Iraq. In an area inhabited by Kurdish tribesman who are Moslems and members of the conservative Sunni sect, Ralph Solecki, a human paleontologist at the Smithsonian Institution, found in the 1950s several skeletons that would prove to be of major importance in the study of human evolution.

The account of the discovery of the cave and Solecki's initial reaction is quite interesting.

Finally, on July 10, having heard so much about the cave of Shanidar from our friend, the district governor . . . I was determined to ask the authorities for permission to investigate it. . . . We bought plenty of gasoline, got our boots shined, had a good lunch, and left about 2:00 PM, the time when all sensible Iraqis retire for their midday rest. . . . It was an attractive, wild country, wilder than any we had ever seen before. . . . The craggy, precipitous heights of the Baradost Mountain towered about us all the way. . . . The people who inhabited the area were the Surchi tribesmen who had been described as the wildest type of Kurd, rough, savage creatures more like bears than human beings. . . . We finally arrived at the village of Shanidar at 5 p.m. The Shanidar valley lies at an elevation of about 1,400 feet . . . the maximum elevation of the Baradost Mountain was about 6,800 feet. It was the middle of the summer and the heat was intense. We kept asking the policeman who was guiding us, how much farther we had to walk in the heat. The reply was always, "Not far, just a little more." Finally, we rounded the nose of the mountain, and there before us, some five hundred feet away, was the cave. It lay a bit more above us and the trail dipped and then swung to the right, and then went left upward to the cave. It presented a huge triangular orifice, with several lens-shaped fissures on the right side. . . . I literally gawked at it for a few minutes. It was certainly the best one we had seen in all of our travels in northern Iraq. I must admit I felt a twinge of excitement as I approached the cave. . . . The roof of the cave was blackened as if from smoke. There was evidence of recent habitation in the cave and I counted a number of twig enclosures. The floor of the cave was thick with dung from cattle, sheep and goats. There must have been at least ten to fifteen feet of deposit of debris at the mouth of the cave. The cave could have housed as many as seventy to eighty people.[46]

Remains from at least nine individuals have been found in the cave, but one skeleton is of particular importance. The skeleton is called Shanidar I and it is an adult male who died as a result of the collapse of the cave ceiling. The skeleton had an underdeveloped right shoulder blade, collarbone and upper right arm. Shanidar I had been a severely disabled individual. There was skeletal damage to his right eye so severe that it would have been impossible to see. The specimen had an atrophied uppermost portion of the right humerus and the lower portion as well. All of the remaining bones of the right forearm were also missing. The lower portion of the right arm was amputated at or slightly above the elbow, an operation that Shanidar I survived. Paleontologists estimate that he lived to a fairly old age (30–35) in Neandertal terms. Interestingly, pollen remains in another burial in the cave indicated the presence of *Ephedra altissima*.[47] The active alkaloid in *Ephedra* is ephedrine, as well as pseudoephedrine. These substances act like adrenaline on the sympathetic nervous system (causes increases in blood pressure and heart rate, breathing increases, levels of blood glucose increase) and could have been used for treatment of shock. In addition, the left ankle of Shanidar I was abnormally developed, suggesting an injury to the lower leg early in life. The right ankle was arthritic possibly from its weight-bearing role. It would have been impossible for this Neandertal to survive without aid.[48]

He obviously could not hunt for himself, let alone walk without assistance. If, as paleoanthropologists believe, this individual was being cared for by the rest of the members of his clan, what contribution would have made him so valuable? It is not clear, but the wear pattern on his teeth suggests that he used his teeth as a tool. Among modern human populations, Eskimos are known for extensive use of their anterior teeth as tools, and, while there are differences between Shanidar and the Eskimo sample, the general conclusion about the use of the teeth is confirmed.

In addition to the material from Shanidar, the skeleton of an achondroplastic dwarf has been found in an Upper Paleolithic site in Italy.[49] It has been argued that this is another example of early humans giving social support to group members who, without support, would not likely have been able to take care of themselves. Neandertals provided for and took care of their elderly and sick individuals.[50] However, there is some disagreement with this position.[51] Taken on the whole, these observations suggest that by the time of the Neandertals in central and western Europe, fifty thousand or so years ago, our ancestors had diverged from the widespread laissez-faire attitude of nonhumans toward injured or debilitated conspecifics and adopted a pattern of helping that sets humans apart from other species.

In order to fully appreciate the fundamental nature of giving aid and social support to others, it is necessary to take a brief look at the history of the welfare system.[52] By the Middle Ages in Europe, the infrastructure was established for dispensing aid to those in need. Donations from churches as well as from local feudal lords and other wealthy individuals supported hospitals, orphanages, and almshouses (poorhouses).[53] The first of the extensive state-supported efforts was the Elizabethan Poor Law of 1601, which attempted to classify dependants and provide special treatment for each group on the local (parish) level.

Two classes of poor were identified. Those deemed worthy of support ("deserving poor") were those who were unable to work—primarily the disabled, blind, and elderly. They were eligible for cash or other forms of assistance. The able-bodied unemployed were labeled the "undeserving poor" but were nevertheless provided with what amounted to public-service employment. Such government-funded work was known as indoor relief, because it was usually done inside large public facilities called workhouses.

The Poor Laws made the local governments responsible for the administration of this type of aid, and, in order to oversee its dispensation, recipients were discouraged from moving from one local political entity to another. The Poor Laws were a magnet for controversy from their inception. People feared that individuals would come to rely on assistance and would not be motivated to work, or, if they did work, they might not be inclined to save any of their earnings. By the late 1700s, formal work programs had been established, and relief recipients were assigned to work in farms or businesses with their salary being partially offset by the government. By the late 1830s, workhouses were established where those that were eligible for aid but could not find work were employed in doing a variety of menial tasks. In fact, the idea was to make the workhouses such an unpleasant environment that people would find other employment as soon as possible.

By the late nineteenth century, programs comparable to the Poor Laws had spread to other parts of Europe. Social programs were instituted in Germany, France, Belgium, and Sweden that provided insurance, allowances for children, universal health coverage, and other programs that offered income and other essential supports for individuals if they were disabled or lost their jobs. The first modern government-supported social welfare program for broad groups of people, not just the poor, was implemented in Germany in 1883. Legislation provided health insurance for workers, while subsequent legislation introduced compulsory accident insurance and retirement pensions. Over the next fifty years, spurred by socialist theory and the increasing power of organized labor, state-supported social welfare programs grew rapidly. By the 1930s, most of the world's industrial nations had some type of social welfare program.

By the early eighteenth century, the British colonies in North America had adopted their own version of the Poor Laws of Great Britain. One of the main differences between the two systems was that, in the colonies, the poor could be auctioned off to bidders who could use them as workers. By the early nineteenth century, a reform movement spread through much of the United States that discouraged direct aid to poor individuals but instead placed them in workhouses. This was seen as a way of discouraging slovenliness by instilling the Protestant work ethic.

By the late nineteenth century, this movement had transformed itself into what was called "the scientific charity movement." This was a plan of reform that was intended to foster independence among the poor by social casework, a practice that involved visitations by a caseworker to instruct the poor in issues of morality and instill in them the Protestant work ethic. The end of the Civil War brought the enactment of legislation that would ultimately offer benefits to any veteran of the Civil War. Congress also passed bills that would provide mothers' pensions primarily for poor widows, thus affirming the idea that motherhood was a full-time occupation.

The modern welfare system arose with the Great Depression. Widespread bank failure, an inability of industry to respond to the market after World War I, and widespread layoffs were both the cause and the effect of the stock market crash. A serious crash in 1929 and a steady decline from mid-1930 until 1933 brought the American economy to its knees. While President Hoover made efforts to end the Depression, he was opposed to providing direct federal relief programs. This opposition convinced voters that he was not sympathetic to the plight of the working man and led to his failure to be reelected. It was Franklin Delano Roosevelt who defeated Hoover in the 1932 election.

In his inaugural address, Roosevelt proclaimed that "the only thing we have to fear is fear itself" and promised "to wage a war against the emergency." The first hundred days of Roosevelt's presidency would be the opening battle. Roosevelt's New Deal consisted of a flurry of legislation that created the "alphabet agencies"—FDIC (Federal Bank Deposit Insurance Corporation), CCC (Civilian Conservation Corps), TVA (Tennessee Valley Authority), AAA (Agricultural Adjustment Administration), FCA (Farm Credit Association), HOLC (Home Owners' Loan Corporation), FHA (Federal Housing Administration), SEC (Securities and Exchange Commission), FCC (Federal Communications Commission), PWA (Public Works Administration), WPA (Works Progress Administration), REA (Rural Electrification Association), NYA (National Youth Administrations), and NRA (National Recovery Administration). It is sometimes said that Roosevelt's first one hundred days were conservative and sympathetic toward business and the second one hundred days were a coalition against business.

The Social Security Act of 1935 provided for federally funded financial assistance to the elderly, the blind, and dependent children. Coverage was initially limited, and domestics, agricultural workers, and people working in businesses of fewer than eight employees were excluded. The act signaled a fundamental change in the direction and outlook of the United States, joining other industrialized nations in providing some social insurance against the perturbations of modern economic life. Subsequent amendments broadened the act in terms of coverage provided and eligibility; included was the provision for medical insurance to the aged under the Medicare program and to low-income families under the Medicaid program, both enacted in 1965.

Huey Pierce Long

Any discussion of the historical dimensions of the welfare program must include some mention of Huey Pierce Long. He was born 10 August 1893 in

Winnfield, Winn Parish, Louisiana, into a middle-class landowning family. He was educated at the University of Oklahoma and Tulane University but did not graduate from either. Long dropped out of Tulane Law School at age twenty-one but passed the bar exam anyway, in 1915, and then practiced law in Shreveport for ten years. During this time, he took on compensation suits for injured workers against the concentrated wealth of big corporations and generally was successful. He was elected to serve on the Louisiana Railroad Commission (later called the Public Service Commission) in 1918 and remained on the commission until 1926, serving as chairman for five years. He was elected governor of Louisiana in 1928 at age thirty-five, after campaigning on a slogan from William Jennings Bryan: "every man a king, but no one wears a crown." Long was the first southern politician to use radio as a primary campaign device.

Long got the nickname "Kingfish" from the character on the *Amos 'n' Andy* radio program. Set in Harlem, *Amos 'n' Andy* centered on the activities of George Stevens, a conniving character who was always looking for a way to make a fast buck. As head of the Mystic Knights of the Sea Lodge, he held the position of "Kingfish." In the story, Kingfish got most of the lodge brothers involved in his schemes. While there is some controversy over whether Long named himself or the moniker was given to him by one of his associates, Long is said to have answered the telephone, "This is the Kingfish."

During his tenure as governor, he was responsible for paving over three thousand miles of roads, constructing 111 bridges, building the New Orleans airport, and beginning construction on a medical school at Louisiana State University in Baton Rouge. All of these infrastructure improvements were largely paid for with a two-cent increase in the sales tax on gasoline. Long established night schools and was able to reduce illiteracy among blacks from 38 percent to 23 percent. He started the first prisoner rehabilitation program at the Angola Prison. And he provided free textbooks to all students, paid for with a severance tax levied on oil and gas companies.[54]

From the initial passage of the Social Security Act in 1935 to the additions of Medicaid and Medicare in 1965, public assistance programs have increasingly come under governmental control. By the early 1990s, the Clinton administration approved changes in many states' welfare systems, including work requirements in exchange for benefits (so-called workfare) as well as rigidly enforced time limits on receipt of assistance.

In 1996, President Clinton signed a bill enacting the most sweeping changes in social welfare policy since the New Deal. In general, the bill, which sought to end long-term dependence on welfare programs, represented a sharp reduction in many benefits as well as a reversal of previous welfare policy, shifting some of the federal government's role to the states. Among the bill's major provisions was the requirement that 25 percent of the population then on welfare should be working or training for work by 1997 (a goal that was reached in most states) and that 50 percent should do so by 2002. States were granted lump sums to run their own welfare-to-work programs. The federal guarantee of cash assistance for poor children was done away with. The limitation of lifetime welfare benefits to five years (with hardship exemptions for some) was imposed. Every head of a welfare family had to go to work within two years of receiving benefits or lose them. Finally, stricter eligibility standards for the Supplemental Security Income program (which excluded many poor disabled children from benefits) were enacted.

In terms of reducing the welfare rolls, the bill initially proved successful; in 1999, there were fewer welfare recipients than there had been in thirty years. Most states also reported a surplus of federal welfare funds. Those funds, which remain by law fixed for five years, provided an unforeseen benefit for the states, enabling some to increase social welfare spending. Skeptics fear, however, that the long-term effects of the reform will be to force welfare recipients into jobs that do not pay a living wage and to burden private charitable organizations with costs the government is no longer willing to bear.

One of the problems with any system that redistributes resources is that it is vulnerable to exploitation. A remarkable aspect of human cognitive adaptations is the ability to manipulate the system. Whether it is a simple trade of goods for services or something much more sophisticated, some individuals are likely to try to gain a competitive advantage at the expense of others.

Altruism, Cheating, and Itta Bena, Mississippi

Where groups are small and individual identity is well known, people are predicted to act more cooperatively. If you have a high probability of meeting someone in the future with whom you have interacted before, you have a significantly increased probability of engaging in an altruistic interaction. Remember the example I discussed earlier of dining in a strange town. Imagine driving in a large metropolitan area as compared to driving in a rural village. One of the most intimidating places in the world to drive is Mexico City at rush hour. The population in the greater metropolitan area exceeds twenty-two million people in approximately 350 distinct neighborhoods, covering over two hundred square miles with a diameter of forty miles. The population of Mexico City itself is over nine million people with five million cars. Rush hour is chaotic, with drivers completely ignoring lanes painted on the streets. Sidewalks become fair game for irate commuters, and the noise level exceeds that of a rock concert.

Vendors race from car to car peddling newspapers, clothing, watches, and jewelry. Most unsettling perhaps are the "los niños de la calle," street kids who dress up in clown suits, who juggle or breath fire in an effort to get a few pesos while cars are standing still, before the light changes and the rush begins again. In this kind of environment, tempers flare, voices are raised, epithets about ancestry are yelled at offenders, and, at its worst, drivers resort to physical violence. (See chapter 2 for a discussion of road rage.) Compare this situation to driving in a rural town somewhere in the United States.

The kind of behavior that is commonplace in Mexico City simply would not exist in a thriving metropolis like Itta Bena, Mississippi (population approx. 2,300).[55] I am not suggesting anything about the relative merits of Mexico versus Mississippi, except that in Mexico City you have an almost zero probability of encountering an offending driver a second time. Given the number of drivers in Itta Bena, the probability goes up considerably. Clearly, if you are concerned about your reputation as a driver, you are likely to be much more altruistic and cooperative in Itta Bena.

This example brings up a fundamental problem for the altruist: the chance that he or she will be duped into helping someone who will not reciprocate. This classic problem has been widely discussed in evolutionary biology, economics, and mathematics under the guise of the Prisoner's Dilemma. The Prisoner's Dilemma presents us with a powerful example of how cooperation might be achieved in a population of nonaltruists. Puzzles with this structure were devised and discussed by Merrill Flood and Melvin Dresher in 1950 as part of the Rand Corporation's investigations into game theory. Rand pursued game theory because of possible applications to global nuclear strategy. The title "Prisoner's Dilemma" and the version with prison sentences as payoffs were created by mathematician Albert Tucker, who wanted to make Flood and Dresher's ideas more accessible to an audience of Stanford psychologists. The Prisoner's Dilemma is the classic example of a "nonzero-sum" game in economics, political science, evolutionary biology, and, of course, game theory.[56]

In the Prisoner's Dilemma, you and a friend are picked up by the police and interrogated in separate cells without a chance to communicate with each other. For the purpose of this game, it makes no difference whether you or your friend actually committed a crime. You are both told the same thing:

- If you both confess, you will both get four years in prison.
- If neither of you confesses, the police will be able to pin part of the crime on each of you, and you'll both get two years.
- If one of you confesses but the other doesn't, the confessor will make a deal with the police and will go free while the other one goes to jail for five years.

At first glance, the correct strategy appears obvious. No matter what your friend does, you'll be better off "defecting" (confessing). Sadly, from your perspective, your friend realizes this as well, so you both end up getting four years. Ironically, if you had both "cooperated" (refused to confess), you would both be much better off. If you play repeatedly, the goal is to figure out your friend's strategy and use it to minimize your total jail time. Your friend will be doing the same thing. Remember, the object of the game is not to harm your friend but to minimize your own jail time. If this means ruthlessly exploiting your friend's generosity, then it is in your best interest to do so; likewise, if it means helping your friend by cooperating, then by all means you should do so.

The "dilemma" faced by prisoners is that, whatever the other does, each is better off confessing than remaining silent. But the outcome obtained when both confess is worse for each than the outcome they would have obtained had both remained silent. A common view is that the puzzle illustrates a conflict between individual and group rationality. A group whose members pursue rational self-interest may all end up worse off than a group whose members act contrary to rational self-interest, and of course this goes against the underlying assumption about human decision making.

We can imagine a world filled with examples of this kind of conflict. What is interesting for humans is that we routinely engage in cooperative behavior, so how relevant is this game theory approach to understanding our cooperative behavior? Well, it offers a way to think about the alternative paths by which individuals can arrive at individual gain, not necessarily at the expense of another.

It is important to see how game theory is used in mediating conflicts on an everyday basis, as well as how it was used historically. Robert Axelrod, a professor of political science and public policy at the University of Michigan, outlined a classic case of a "live-and-let-live" strategy played out during World War I.[57] Trench warfare in Europe was one of the bloodiest and most gruesome types of conflict in history. The major feature of trench warfare was the erection of a barrier that could not be overrun by the enemy but could be defended economically.

A classic example of this type of warfare was carried on along a five-hundred-mile line through France and Belgium representing the battle line between English and German armies. The trenches were dug to protect troops from deadly machine gun fire. Actually two parallel trenches were dug by each side. The firing trench was the site of primary military activity, but there was a cover trench that was a backup in case the firing trench was overrun. Trenches were six to eight feet deep, and soldiers lived in the secondary trenches. A network of trenches connected the front lines with supply areas away from the line of fire.

Between the trenches of the opposing forces was a no-man's-land that

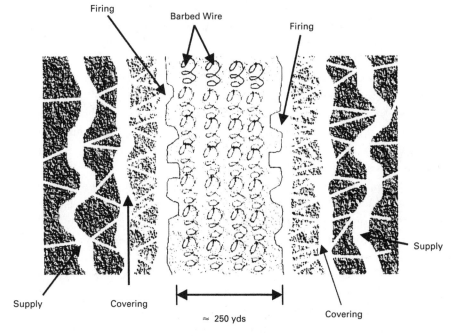

Fig. 6-1. Diagram of hypothetical battle lines in Europe during World War I. The main firing trenches were separated from the opposition forces by a "no-man's-land." The covering trenches were the main line of defense and served as support for those in the firing trenches. Behind the covering trenches were one or more supply trenches. All guns, ammunition, food, clothing, and other supplies had to be moved through the supply trenches. Major movements outside the trenches were conducted in darkness. Main trenches were connected by smaller trenches. © E. O. Smith, MMI.

was defined by barbed wire fences. The width of no-man's-land varied considerably along the western front. The average distance in most sectors was about 250 yards. At its minimum, no-man's-land measured no more than 50 yards (Guillemont). At its maximum (Cambrai), it was over 500 yards. No-man's-land contained a considerable amount of barbed wire fencing. In the areas most likely to be attacked, there were as many as ten rows of barbed wire just before the frontline trenches. In some places, the barbed wire area was more than a hundred feet wide.

By 1914, trenches extended from the North Sea to the Swiss border. Germany had dug in first and had taken the best locations for its fortifications. The Allies were relegated to the lower ground and often were digging trenches in ground only two to three feet above sea level. Consequently, many Allied soldiers lived in standing water or mud.

Soldiers developed a particularly interesting code of conduct in the midst of so much death and destruction. So long as they were not poised for an all-out assault, opposing soldiers tacitly agreed that they might as well make life as comfortable as possible for each other. Certain informal patterns of conduct developed. Artillery was fired only at particular times of day. Since military command had made it clear that a certain number of shells had to be fired every day in order to make life miserable for the enemy, it was agreed that the shelling would be done, but it would be at times when the opposition was *not* having dinner. Meals were a terribly important part of life in the trenches on the western front, and soldiers looked forward to them as a high point in the day.

Interestingly, a kind of Golden Rule developed in the trenches. If you had to fire an artillery barrage, then it should not be done at mealtimes. Moreover, you would not fire at a position that was likely to hurt many of the enemy. You did not fire at their trenches. In this way, soldiers did their duty and fired off the required number of shells. They demonstrated that they were hostile and ready for combat, and so fulfilled the letter, if not the spirit, of the rules of engagement.

Axelrod shows that this type of trench warfare is an example of the "Iterated Prisoner's Dilemma."[58] In a given locality, the players can be thought of as the local units facing each other across the "no-man's-land." The choices at any time are shoot to kill, or shoot to avoid doing damage. Weakening the enemy is important so that, if an all-out battle is ordered, your opponent's strength has been reduced. In the short run, it is better to do damage, whether or not your opposition is retaliating. The distinctive feature of trench warfare in World War I was that the same troops faced each other on a daily basis for an extended period of time. This changed the game from a classic "One-Move Prisoner's Dilemma" to an "Iterated Prisoner's Dilemma," in which conditional strategies are possible. It would be in the best interest of both parties to refrain from defecting, but to defect immediately if the opposition did so.

The point of this example is to demonstrate how cooperative behavior might evolve in the most unusual of circumstances. Additionally, this example also shows how cheating can be kept in check. We know that one of the most serious blocks to the evolution of cooperative behavior is vulnerability to cheaters, so in our World War I example, retaliating "cheaters," or defectors who shot to kill, were dealt with in like manner. No gain to cheating was allowed under these circumstances. Reciprocity turns out to be a fundamental dimension of human behavior. It very well may be one of the true hallmarks of our humanity.[59]

Charity, Welfare, and the Future

In the United States, there has developed an impressive system for the redistribution of wealth, with a primary goal to help those who otherwise would not be able to survive. There is, however, widespread dissatisfaction on the part of both taxpayers and recipients that the system does not work, that it treats some better than others, overlooks some needy people completely, provides resources to those that really do not need them, and is so vulnerable to abuse that it is a disgrace. Those who support the system with money complain that taxes are too high and there is mismanagement and abuse in the welfare system. They cite cases of individual as well as corporate abuse. Those receiving aid complain that it is not enough to meet their needs, or that the bureaucratic system is so cumbersome that it is not easily accessible. Those who administer the system complain that they are overworked and underappreciated.

So what does the evolutionary perspective offer us as a way of designing a better system for redistribution of wealth and assistance to the needy? Some would say nothing, that evolution in the biological sense is irrelevant to questions of social policy. On the other hand, much social policy has been enacted without any theoretical underpinning to guide its implementation. If we had some general guidelines about what to expect, it is at least possible that one could devise a system that would meet the larger charitable objectives. Evolutionary theory is able to generate predictions about human behavior based on experience with a variety of animal and bird species. It seems extraordinarily arrogant to think that human behavior is so far removed from that of nonhumans that we cannot learn anything about ourselves by observation of other species.

We should recognize that our attitudes toward welfare and entitlement programs are the product of our fundamental view of the nature of humanity emerging in a complex early environment. Whether we are basically good and have been corrupted by forces of society, or if we are fundamentally selfish and must be coerced into acting cooperatively, is an important issue that underpins many of our attitudes and much of our behavior. It is clear that there are people who have become well known for their genuine kindness and generosity, even to the point of self-destruction. To be sure, this is curious behavior, and behavior that sets us apart from other social species.

The problem that confronts us is the routine commission of the "naturalistic fallacy," which, crudely put, states that what is, is what ought to be. We are confronted with the view of some that a hierarchical distribution of resources is an inevitable part of society that we can do little about and, in fact, differential control of resources is the stuff that glues society together. Adoption of this view sets up a particular set of expectations about the welfare system and aid to the needy as we know it. On the other hand, there are a substantial number who feel that what ought to be, should be. Some sense of justice and fairness is a goal of human interaction, and the means to those ends should be implemented in society.

Our study of animal behavior offers examples of how to encourage altruistic or cooperative behavior. Using the calculus of evolutionary biology, it is easy to imagine how acting altruistically to someone who is kin would benefit us. The vast majority of cooperative behavior in animals is not the result of higher cognitive abilities, or the specific outcome of the evolution of ethics or morals, but is simply the outcome of basically genetically selfish behavior. To the extent that we can be convinced that people are our kin, we are predicted to be more likely to give aid. It should come as no surprise then that charitable organizations figured this out long ago, along with the government and virtually every special-interest group seeking financial assistance. Marketing phrases that capture the idea of the unity of mankind, our fundamental similarity no matter the color of our skin, or the equality of all, are used to encourage charitable behavior.

On the other hand, without these powerful messages, humans seem to be capable of extraordinary feats of self-sacrifice and compassion that are unexplainable even by the most sophisticated of the delayed cost-benefit models. The regularity with which disasters elicit a kind of self-sacrificing altruism is well documented. What possesses people to put themselves in harm's way, when the likely outcome is their own death? This is a question that we are only beginning to understand, and we have yet to derive a satisfactory answer.

All this is to say that our welfare system was born of a concern for others, and it has a long and distinguished history that has been examined by an array of scholars, theologians, psychologists, psychiatrists, and clinical practitioners. If we could develop a more complete understanding of the cooperative dimension of human behavior, it might have a profound impact on our ability to live in this increasingly challenging world.

Given what we know about our culture and biology, and the comparative data we have from other species on altruism, it is possible to offer some ideas for the future. The ideas offered below are intended to serve as a starting point for discussion.

- Given the propensity for humans to act in their own self-interest, programs that help others would be most successful if altruistic donors could be convinced that there was a probability of some future reciprocity (tax breaks, preferred seating at public events, reserved parking places, etc.).
- People are more likely to engage in altruistic behavior when the costs of cheating are clearly spelled out and there is strong likelihood of detection. Charity toward individuals who are really in need is likely to be successful, but one must be on guard against those who exaggerate their need in order to gain an advantage. Penalties for cheating might include exclusion from any future aid.
- One of the things that is an important part of the World War I trench warfare example is that the environment in which cooperation evolved was extreme. We are much more likely to develop cooperative behavior when the costs of not doing so are high. So it follows that likely areas where cooperative behavior can be fostered are those where cooperation and altruism will reap large benefits and failure to do so would be catastrophic.
- A greater willingness to engage in charitable behavior might be favored if resources were used on a local level. When givers and receivers of aid know each other over a long period of time, the likelihood of cheating is diminished. It is much easier to imagine someone abusing a charitable gift that comes anonymously or, even worse, from a central repository for resources than one from a member of the same community. Guilt is a powerful motivator of behavior

under the right set of ecological and social conditions, and one could argue that intimate, face-to-face interactions increase the probability for the expression of guilt in the face of exploitation.

- Not only is abuse decreased when charity is given and received on a face-to-face basis; the willingness of people to be altruistic increases. We know that people are more likely to engage in charitable behavior if the recipient of the aid is known to the donor. Charitable campaigns conducted on a national level have to overcome this lack of person-to-person contact through elaborate marketing campaigns that are aimed at convincing potential donors of the reality and humanity of the recipients. It is not surprising that charitable organizations rely on the appeal of children to induce donors. How many organizations have a "poster child" to represent them? Don't get me wrong. I am not saying that this is necessarily bad, but it is a type of psychological manipulation. One of the most successful of the nonprofit charitable organizations is Habitat for Humanity. It is an organization that builds simple, decent, affordable housing for people around the world. Since its inception, Habitat for Humanity has built over 100,000 houses, sheltering more than 500,000 people worldwide. Its success is based on face-to-face interactions of people helping others to construct a house.

Americans are concerned about the continuation of entitlement programs to which many have contributed their entire working career. There is a lingering fear among the "baby boomers" that Social Security funds will be exhausted before they receive any return on their investment. Reports of gloom and doom related to the solvency of the Social Security Administration (SSA) are nothing new, but recent projections of the SSA running at a deficit in the next two decades had not previously been heard. As of 2000, the Social Security trust reserves were projected to be adequately financed until 2037,[60] but we all know that economic forecasting is often more like an art than a science. Social Security is but one part of a vast collection of entitlement programs that together are a major proportion of the total budget of the United States.[61] While there are many things to recommend the system of entitlement programs in the United States, there are still areas that need considerable work in the formation of policies that will actually benefit the poor. Too often, policy is made without any explicitly stated theoretical background, but nevertheless with an agenda dictated by our core assumptions about humanity. Once again the fundamental question arises: who is right, Hobbes or Locke?

It is unlikely that we will have evidence to completely satisfy all members on either side of the debate. What we can do, however, is to use basic Darwinian evolutionary theory to make some predictions about how humans ought to behave when they are relieved, to some extent, of the necessity of obtaining basic resources. Moreover, we should be able to design systems of dispensation of assistance that are more consistent with basic human predispositions than many of the ones in place today. Recognizing that we do not come to any situation with a completely blank slate is a large part of designing political systems that have a reasonable chance of achieving their goals.

Richard Dawkins, an evolutionary biologist at Oxford University, in his famous book *The Selfish Gene*, noted that we are the only species that has the ability to overcome the tyranny of the genes. We are the only species to be able to act in a truly altruistic way, but it is clear that it will take some work. One of the major advantages of an evolutionary perspective is, not that it will provide specific outlines for programs, but that it can offer some guiding principles that may help us design programs that will help those in need. Like the chess-playing computer, if we understand the basic rules of the game, we can better orchestrate a successful outcome.

Notes

1. Intersection of Biology and Culture

1. Edward Burnett Tylor, 1871, *Primitive Culture: Researches into the Development of Mythology, Philosophy, Religion, Art, and Custom* (J. Murray, London).

2. Sarah Hrdy in her book (1999) *Mother Nature: A History of Mothers, Infants, and Natural Selection* (Pantheon Books, New York) notes that opium may be smeared on mothers' nipples or infants may be poisoned directly.

3. William Shakespeare, 1591, in *King Henry the Sixth*, part 3, ii, 104.

4. Mead T. Caine, 1977, The economic activities of children in a village in Bangladesh, *Population and Development Review* 13 (3), 201–227.

5. We tend to think of investment in offspring in terms of resources given to one that we cannot give to another. However, an animal behaviorist, James Wittenberger, points out that certain kinds of parental investment can be shared; for example, a parent can guard a nest having more than one offspring as efficiently as one with a singleton. The rub comes, not with defending the offspring, but providing food. In this case, the protection is shareable, while food is not. James F. Wittenberger, 1981, *Animal Social Behavior* (Duxbury Press, Boston).

6. N. Tinbergen, 1963, On aims and methods of ethology, *Zeitschrift für Psychologie* 20, 410–433.

7. See the recent find in Kenya that pushes bipedalism back to six million years ago: B. Senut et al., 2001, First hominid from the Miocene (Lukeino Formation, Kenya), *Comptes Rendus de L'académie des Sciences, Paris* 332, 137–144.

8. *Striding*, as opposed to hopping bipedal locomotion, has evolved in two vertebrate lineages, archosaurs and primates, and is represented today by humans and birds, while *saltatory* bipedalism is found among macropod marsupials, some rodents, and some birds (M. Djawdan and T. Garland, 1988, Maximal running speeds of bipedal and quadrupedal rodents, *Journal of Mammalogy* 69 (4), 765–772). Interestingly, and not surprisingly, anthropologists concerned with the evolution of human locomotion have largely overlooked birds or other vertebrates as taxonomic groups that might provide important information on gait and limb usage, potentially furthering our understanding of human bipedalism. In fact, birds constitute a group of species that exhibits many more forms that are adapted to a striding bipedal locomotion as compared to primates. S. M. Gatesy and A. A. Biewener (1991, Bipedal locomotion: effects of speed, size and limb posture in birds and humans, *Journal of Zoology* 224, 127–148), in a careful analysis of the gaits of seven species of ground-dwelling birds (0.045–90.0 kg), found that humans and large birds exhibit striking similarities in stride length and stride frequency in relation to speed with a given gait

but the transition between gaits occurs abruptly in humans: step length, limb excursion angle, and duty factor all fall sharply in the walk-run transition, while birds have a much smoother gait transition in all of the above-mentioned parameters. In fact, the transition from walking to running, while obvious in humans, is not at all conspicuous in the large birds. In summary, humans have a stiff run that favors storage and recovery of energy, whereas birds have a type of run that favors stability. Gatesy and Biewener (Bipedal locomotion) suggest that the differences between human bipedalism and that of bipedal dinosaurs, possessing a large counterbalancing tail, are similar to those between humans and the extant ground-dwelling kangaroo.

9. If we compact the time since the big bang into a calendar year, then Earth was not formed until mid-September, and it was only late September when life originated. The first human ancestors appear on New Year's Eve at 7:30 P.M. or so, with modern humans arriving around 11:40 P.M. The earliest cave paintings in Europe occurred at 11:59, the rise of agriculture at 11:59:20, the Renaissance in Europe, the Ming Dynasty in China, the emergence of the scientific method by 11:59:59, and the widespread development of science and technology, space travel, AIDS, and MTV in the last second before midnight. Carl Sagan, 1977, *The Dragons of Eden: Speculations on the Evolution of Human Intelligence* (Random House, New York).

10. George J. Armelagos, 1987, Biocultural aspects of food choice, in M. Harris and Eric B. Ross, eds., *Food and Evolution: Toward a Theory of Human Food Habits* (Temple University Press, Philadelphia, pp. 579–594); Mark Nathan Cohen, 1989, *Health and the Rise of Civilization* (Yale University Press, New Haven); Mark Nathan Cohen and George J. Armelagos eds., 1984, *Paleopathology at the Origins of Agriculture* (Academic Press, New York); Jared M. Diamond, 1987, The worst mistake in the history of the human race, *Discover* 8 (5), 64–66; Marvin Harris and Eric B. Ross, eds., 1987, *Food and Evolution: Toward a Theory of Human Food Habits* (Temple University Press, Philadelphia).

11. S. B. Eaton and M. J. Konner, 1997, Paleolithic nutrition revisited: a twelve-year retrospective on its nature and implications, *European Journal of Clinical Nutrition* 51 (4), 207–216; S. B. Eaton and M. Konner, 1985, Paleolithic nutrition: a consideration of its nature and current implications, *New England Journal of Medicine* 312, 283–289; S. B. Eaton et al., 1988, *The Paleolithic Prescription: A Program of Diet and Exercise and a Design for Living* (Harper & Row, New York).

12. U.S. Bureau of the Census, International Database, *http://www.census.gov/ipc/www/worldpop.html*.

13. Clark Wissler and Lucy Wales Kluckhohn, 1966, *Indians of the United States* (Doubleday, Garden City, N.Y.).

14. Glucose, the most abundant of the monosaccharides ($C_6H_{12}O_6$), is the raw fuel of the biosphere and is used as the repeating unit in most of the more complex sugars and starches.

15. The discussion of sugar metabolism and diabetes follows Carl E. Rischer and Thomas A. Easton, 1992, *Focus on Human Biology* (HarperCollins, New York).

16. W. P. Newman and R. G. Brodows, 1983, Insulin action during acute starvation: evidence for selective insulin resistance in normal man. *Metabolism: Clinical & Experimental* 32 (6), 590–596.

17. A bottleneck is a severe temporary restriction in the size of a population. A particularly interesting type of bottleneck occurs when a new population is established by a small number of founders, called the founder's effect.

18. One of the most famous Pima Indians in World War II was from the Gila River Reservation. Ira Hamilton Hayes was born 12 January 1923, the son of Joe and Nancy Hayes. Ira joined the Marines at the outset of World War II and was sent to parachutist School at San Diego, California, where he was nicknamed Chief Falling Cloud. By the beginning of 1945, he was part of the American force that attacked the Japanese stronghold of Iwo Jima. On 23 February 1945, Hayes and five others raised the United States flag atop Mount Suribachi on the island of Iwo Jima, signaling the end of Japanese control in the region. Three of the six men were killed while raising the flag, but the heroic act was photographed by Joe Rosenthal, transforming Ira Hayes's life forever.

19. E. Ravussin et al., 1994, Effects of a traditional lifestyle on obesity in Pima Indians, *Diabetes Care* 17 (9), 1067–1074.

20. Today, about 12 to 13 percent of the population is left-handed, but this is in marked contrast to the 0.7 percent of Taiwanese students. When a culture relaxes its sanctions on a particular behavior, we revert to our evolved ways.

21. No wonder video games and computer graphics are so popular. One of the frequent comments on course evaluations from my students is how much they enjoyed the videos that I showed during the semester.

22. Ernst Mayr, 1997, *This Is Biology: The Science of the Living World* (Harvard University Press, Cambridge, Mass.).

23. G. A. Harrison et al., 1988, *Human Biology: An Introduction to Human Evolution, Variation, Growth, and Adaptability* (Oxford University Press, Oxford).

24. R. D. Alexander et al., 1979, Sexual dimorphisms and breeding systems in pinnipeds, ungulates, primates, and humans, in N. A. Chagnon and W. Irons, eds., *Evolutionary Biology and Human Social Behavior: An Anthropological Perspective* (Duxbury Press, North Scituate, Mass.) pp. 402–435.

25. Discussion following M. Ridley, 1996, *Evolution* (Blackwell Science, Cambridge.)

26. Eaton et al., *The Paleolithic Prescription*.

27. The reasons why one sex is in short supply is beyond the scope of this chapter, but for discussion and extensive references see R. L. Trivers and D. E. Willard, 1973, Natural selection of parental ability to vary the sex ratio of offspring, *Science* 179, 90–92, and T. H. Clutton-Brock and G. R. Iason, 1986, Sex ratio variation in mammals, *Quarterly Review of Biology* 61, 339–374.

28. Charles Darwin, 1871, *The Descent of Man, and Selection in Relation to Sex* (John Murray, London).

2. Road Rage, Stress, and Evolution

1. Mike Clary, 2001, Simpson faces road rage charges; law: football hall of famer is arrested on burglary and battery counts in Miami incident. He says he welcomes felony case, *Los Angeles Times,* 10 February, home edition, section A; part 1, p. 3. Simpson was acquitted 24 October 2001, for lack of evidence.

2. Oxford University Press, 1996, *The Oxford Modern English Dictionary* (Oxford University Press, New York).

3. The first three quoted examples are from Jason Vest et al., 1997, Road rage, *U.S. News & World Report*, 122, 24–32.

4. Sebastian Rotella and Chris Kraul, 1995, Tank's driver beset by drugs, money problems, *Los Angeles Times*, 19 May, home edition, part A, p. 1.

5. Brian Mooar and Fern Shen, 1998, 2nd driver files complaint saying Tyson assaulted him, *Washington Post*, 2 September, final edition, metro section B, p. 4.

6. Matthew Joint, 1995, *Road Rage* (The Automobile Association Group Public Policy Road Safety Unit, London).

7. LEX Services is a British-based vehicle marketing services company.

8. T. Harding, 2000, U.K.'s drivers overtake Europe for road rage, *The Daily Telegraph*, 15 May, p. 5.

9. Ricardo Martinez, 1997, Statement of the Honorable Ricardo Martinez, M.D., administrator, National Highway Traffic Safety Administration, U.S. House of Representatives, Subcommittee on Surface Transportation, Washington, D.C.

10. Leon James and Diane Nahl, 2000, *Road Rage and Aggressive Driving: Steering Clear of Highway Warfare* (Prometheus Books, Amherst, N.Y.).

11. John Larson, 1999, *Road Rage to Road-Wise: A Simple Step-by-Step Program to Help You Understand and Curb Road Rage in Yourself and Others* (Forge, New York).

12. Arnold P. Nerenberg, 2001, America's leading authority on "Road Rage" and aggressive driving, *http://www.drnerenberg.com/roadrage.htm*.

13. National Highway Traffic Safety Administration and Federal Highway Administration, 1999, *Summary Report: Aggressive Driving and the Law* (Washington, D.C.).

14. This hypothetical speed is based on the current men's outdoor record for one hundred meters (9.79 seconds) held by Maurice Greene of the United States, set on 16 June 1999 in Athens, Greece, converted to miles per hour.

15. Persistence hunting is a technique that requires hunters to follow prey until it is literally exhausted. This means that game may be pursued for days at a time. Grover S. Krantz, 1968, Brain size and hunting ability in earliest man, *Current Anthropology* 9 (5), 450–451.

16. Archaeological research in Eurasia now appears to push the date for the first horseback riding back to approximately 4000 B.C. (Copper Age). Excavations from Dereivka in the Ukrainian steppes have unearthed horse teeth from this period that show distinct signs of bit wear. This means that humans began riding shortly after domestication—some three thousand years prior to significant horseback riding in the "civilized" Near East. As these people had no written language, were nomadic, and utilized materials that have not survived, little more is known of their early riding efforts. It would be more than three thousand years before their legacy, the Scythian cavalry, would make their presence felt in the "civilized" world, around 670 B.C.

17. Harry Reginald De Silva, 1942, *Why We Have Automobile Accidents* (J. Wiley & Sons, New York).

18. Many bicyclists came to grief, 1896, *New York Times*, 31 May, p. 2.

19. P. Marsh and P. Collett, 1987, The car as a weapon, *ETC.: A Review of General Semantics* 44 (2), 146–151.

20. Annual US Highway Fatalities from 1957, *http://www.publicpurpose. com/hwy-fatal57+.htm*.

21. John Macmillan, 1975, *Deviant Drivers* (Saxon House/Lexington Books, Lexington, Mass.).

22. Vehicular homicides are classified as first degree if the driver: (1) unlawfully overtook or met a school bus; (2) unlawfully failed to stop after a collision; (3) drove recklessly; (4) drove while under the influence of alcohol or drugs; (5) failed to stop for, or otherwise was attempting to flee from, a law enforcement officer; or (6) had previously been declared a habitual violator.

23. Georgia Annotated Criminal Code 16-5-1, 16-5-2, 16-5-3, and 16-5-21.

24. D. B. Rathbone and J. C. Huckabee, 1999, *Controlling Road Rage: A Literature Review and Pilot Study* (AAA Foundation for Traffic Safety, Washington D.C.).

25. If this interpretation were correct, it would be interesting to look at public attitudes toward laws governing the sale and distribution of computer software.

26. Macmillan, *Deviant Drivers*.

27. Ibid.

28. Object to the abuse of automobiling, 1902, *Motor World* 4, 662.

29. James J. Flink, 1975, *The Car Culture* (MIT Press, Cambridge, Mass.).

30. T. Conyngton, 1899, Motor carriages and street paving, *Scientific American* 48 (supplement no. 1226), 196.

31. Meyer Hai Parry, 1968, *Aggression on the Road: A Pilot Study of Behaviour in the Driving Situation* (Tavistock Publications, London).

32. Ibid.

33. Ibid.

34. De Silva, *Why We Have Automobile Accidents*.

35. Ibid.

36. Francis Antony Whitlock, 1971, *Death on the Road: A Study in Social Violence* (Tavistock Publications, London).

37. Marsh and Collett, The car as a weapon.

38. Whitlock, *Death on the Road*.

39. A. R. Hauber, 1980, The social psychology of driving behaviour and the traffic environment: research on aggressive behaviour in traffic, *International Review of Applied Psychology* 29, 461–474.

40. Louis Mizell Inc., 1997, *Aggressive Driving* (AAA Foundation for Traffic Safety, Washington, D.C.).

41. David K. Willis, 1997, *Research on the Problem of Violent Aggressive Driving* (House of Representatives, Washington, D.C.).

42. We have had well-documented cases of parking lot rage, a low-speed form of "road rage" in which drivers cannot resist retaliating when someone dings their car door or they believe they have been wronged in the pursuit of choice parking. A battle over a parking spot can turn into bad words, vandalism, and fisticuffs. One

Tustin, California, woman even admitted getting her nine-year-old nephew to urinate on the door handle of a car whose driver "stole" her spot. Jenifer McKim, 2000, Now at the mall: "parking lot rage"; short-tempered drivers vying for a space end up fighting or defacing cars, *Washington Post*, 20 December, final edition, section A, p. 19.

43. Mizell reports that actor Jack Nicholson thought another driver had cut him off and retaliated by stepping out of his car at a red light, grabbing a nine iron, and repeatedly striking the windshield and roof of the offending vehicle.

44. This does not include the rather aggressive "crash and grab" kind of shoplifting, where an individual simply runs a vehicle into a store to steal merchandise.

45. The riskiness of young males has long been recognized. "Ambitious young male drivers are inclined to allow their 'push' to carry over into their motoring. They assume an 'offensive' attitude toward other drivers." De Silva, *Why We Have Automobile Accidents*.

46. In the remainder, 528 (5.2 percent), the sex of the perpetrator was unknown.

47. Michael Fumento, 1998, "Road rage" versus reality, *The Atlantic Monthly* 282 (2), 12–17.

48. Rathbone and Huckabee, *Controlling Road Rage*.

49. Dominic Connell and Matthew Joint, 1996, *Driver Aggression* (MCIT Road Safety Unit, London), for a survey in Britain. See Y. Nagatsuka, 1989, The current situation of traffic psychology in Japan, *Applied Psychology: An International Review* 38 (4), 423–442, for a survey of Japan; C. G. Hoyos, 1988, Mental load and risk in traffic behaviour, *Ergonomics* 31 (4), 571–584, for Germany; and Dr. Leon James's ongoing comparative survey of aggressive driving, *http://www.soc.hawaii.edu/~leonj/leonj/leonpsy/leon.html*.

50. Tubelike organs called hair follicles that lie in the epidermis manufacture hairs. The hair follicles are shaped like a shallot with the bulbous part the growing end. Hair follicles grow at an angle, and about two-thirds of the way up is a bulge with a strap of smooth muscle attached to it. When these muscles contract, they pull the follicles to a more or less upright position, causing the hair to stand up, piloerection. R. J. Harrison and W. Montagna, 1973, *Man* (Appleton-Century-Crofts, New York).

51. K. Belkic et al., 1998, Occupational profile and cardiac risk: possible mechanisms and implications for professional drivers, *International Journal of Occupational Medicine & Environmental Health* 11 (1), 37–57.

52. Ibid.

53. H. F. Harlow and M. Harlow, 1966, Learning to love, *American Scientist* 54 (3), 244–272; H. Harlow et al., 1971, From thought to therapy: lessons from a primate laboratory, *American Scientist* 59, 538–549; W. T. McKinney et al., 1971, The sad ones, *Psychology Today* (May), 61–68.

54. There are some data relevant to this contention. Psychologists are interested in the effect of television violence on expression of aggressive behavior. Significant socializing agents such as television can influence children's perception of and involvement in violence. Studies have shown to varying degrees, depending on the specific research question and methodology, that there exists a statistically sig-

nificant positive correlation between exposure to television violence and antisocial behavior. See Leo Bogart, 1972, Warning: the surgeon general has determined that TV violence is moderately dangerous to your child's mental health, *Public Opinion Quarterly* 36 (4), 491–521; George Comstock, 1986, Television and film violence, in Steven J. Apter and Arnold P. Goldstein, eds., *Youth Violence: Programs and Prospects* (Pergamon Press, Oxford) pp. 178–218; Richard B. Felson, 1996, Mass media effects on violent behavior, *Annual Review of Sociology* 22, 103–128; Amanda E. Pennell and Kevin D. Browne, 1999, Film violence and young offenders, *Aggression & Violent Behavior* 4 (1), 13–28; Waheeda Tabassam, 1994, A study of causal factors of antisocial behaviour in school children, *Bangladesh Journal of Psychology* 14, 69–75. Along the same lines of reasoning, it has been suggested that video games might also have an effect on the expression of antisocial, aggressive behavior. The available literature, while less compelling than that for television, is consistent with the significant positive association. O. Weigman and E.G.M. van Schie, 1998, Video game playing and its relations with aggressive and prosocial behavior, *British Journal of Social Psychology* 37, 367–378. In a study of the viewing preferences of "bad" drivers (those convicted of a moving violation during the three-year period of the study and who were assigned to a county driving school in the state of Michigan) and "good" drivers (individuals who had not been convicted of a moving violation or had not been responsible for an accident during the same period), the hypothesis was supported that "bad" drivers have a greater preference for viewing violent programming on television than "good" drivers. The author cautions, however, that these results do not prove a cause-and-effect relationship. See J. R. Smith, 1969, Television violence and driving behavior, *Educational Broadcasting Review* 3, 23–28. In another study of driving behavior depicted on television, a content analysis of 784 driving scenes in 114 programs revealed that, in less than 3 percent of the scenes, there were immediate legal penalties imposed on bad drivers. Moreover, the greatest numbers of irregular driving acts were depicted in action/drama programs. It is suggested that viewers who watch action/drama shows most frequently are more likely to accept irregular driving behavior as normal, believe that speeding is not serious and is unlikely to result in any penalty, regard seat belts as unnecessary, believe that irregular driving is especially acceptable on rural roads and on hilly terrain, and believe that young people (in their twenties) do more irregular driving than any other age group. See B. S. Greenberg and C. K. Atkin, 1983, The portrayal of driving on television, 1975–1980, *Journal of Communication* 33 (2), 44–55.

55. Larson, *Road Rage to Road-Wise*.

56. Ibid.

57. P. A. Ellison et al., 1995, Anonymity and aggressive driving behavior: a field study, *Journal of Social Behavior and Personality* 10 (1), 265–272; Patricia Ellison-Potter et al., 2001, The effects of trait driving anger, anonymity, and aggressive stimuli on aggressive driving behavior, *Journal of Applied Social Psychology* 31 (2), 431–443.

58. Vest et al., Road rage.

59. Rathbone and Huckabee, *Controlling Road Rage*.

60. David Schrank et al., 1999, *The 1999 Annual Mobility Report* (Texas A&M University, College Station, Tex.).

61. While an RCI of less than one may be impossible, it is truly the dream of commuters. Imagine getting to work sooner when the road is crowded, than when it is less congested. No longer would heavy traffic pose an obstacle to going to concerts, sporting events, on leisure travel, and so on.

62. See Timo Lajunen et al., 1999, Does traffic congestion increase driver aggression? *Transportation Research Part F: Traffic Psychology & Behaviour* 2 (4), 225–236.

63. A. S. Hakkert et al., 2001, The evaluation of effects on drivers' behavior and accidents of concentrated general enforcement on interurban roads in Israel, *Accident Analysis & Prevention* 33 (1), 43–63.

64. Michael G. Flaherty, 1999, *A Watched Pot: How We Experience Time* (New York University Press, New York); Bettina L. Knapp, 1997, Beckett's *That Time:* exile and "that double-headed monster . . . time," *Journal of Melanie Klein & Object Relations* 15 (3), 493–511; Mary Mead Lofy, 2000, A matter of time: power, control, and meaning in people's everyday experience of time. *Dissertation Abstracts International, A* (Humanities and Social Sciences) 60 (12-A), 4628.

65. Other movie examples of "faster is better" include *Mad Max, Death Race 2000, Bullet, Downhill Racer, Grand Prix, Thunder Road, Days of Thunder, Le Mans, Gone in 60 Seconds, Driven, Born to Race, The Gumball Rally, Hell on Wheels*, and the *Roaring Road* (a silent film with an original organ score, starring Wallace Reid, Ann Little, and Theodore Roberts).

66. Vest et al., Road rage.

67. Joel Garreau, 1991, *Edge City: Life on the New Frontier* (Doubleday, New York).

68. There is some disagreement about the attribution of this quote. Some have credited Red Sanders, Vanderbilt University football coach, with the comment. See S. Platt, ed., 1992, *Respectfully Quoted: A Dictionary of Quotations* (Barnes & Noble, New York).

69. Fumento, "Road rage" versus reality.

70. Nerenberg is the author of a plethora of self-help books and materials, including: (1) *Calming Road Rage*, a two-minute tape that is guaranteed to calm you when you play it during your road rage; (2) *Overcoming Road Rage*, a ten-step compassion program to understand and prevent road rage; (3) *The Road Rage Song*, a song of road rage and tragedy; and (4) *Life's Too Short for Road Rage*, cartoon illustrations of forty-nine ways to become an angry driver.

71. Robert Ardrey, 1966, *The Territorial Imperative: A Personal Inquiry into the Animal Origins of Property and Nations* (Atheneum, New York); Michael Chisholm and David Marshall Smith, 1990, *Shared Space, Divided Space: Essays on Conflict and Territorial Organization* (Unwin Hyman, London); R. Dyson Hudson and E. A. Smith, 1978, Human territoriality: an ecological reassessment, *American Anthropologist* 80, 21–41; Julian J. Edney, 1974, Human territoriality, *Psychological Bulletin* 81 (12), 959–975; Robert David Sack, 1986, *Human Territoriality: Its Theory and History* (Cambridge University Press, Cambridge); Pierre L. Van den Berghe, 1977, Territorial behavior in a natural human group, *Social Science Information* 16 (3 supplement 4), 419–430.

72. J. Alcock, 1998, *Animal Behavior* (Sinauer, Sunderland).

73. Joint, *Road Rage*.

74. From an evolutionary perspective, individuals who were able to communicate something about their intentions in a way that did not involve physical contact might have had an advantage. Imagine that you could settle a dispute over a critical resource by the use of nonverbal communication. Individuals who did not have as well-developed communication skills as you would have had to resort to fighting to settle disputes. No matter how primitive the communication might have been, so long as the message was communicated, then physical violence could have been avoided. Hence, those individuals with even the most rudimentary signal generation and detection hardware would have had an evolutionary advantage over individuals who had to risk potential injury in a fight.

75. The V-for-victory sign is associated with Winston Churchill, who took it up in early 1941 and used it as a sign of the anti-Nazi movement. Unfortunately, not all cultures have the same meaning for this common gesture. If the hand is turned so that the palm is facing away from the signaler, the meaning is insulting. So, depending upon which way one's hand is turned, you can signal either victory or peace, or something about one's parentage. See Desmond Morris et al., 1979, *Gestures* (Stein and Day, New York).

76. Life's too short for chess—Henry James Byron. Chess is mental torture—Gary Kasparov.

77. Fédération Internationale des Échecs is one of the two governing bodies for professional chess. The federation, organized in 1924 in Paris, developed championship titles and a system for recognizing top players by arithmetic rating and by titles based on tournament performance. The highest title, after world champion, is international grand master, of which there are now more than four hundred in the world.

78. Computers began to compete against humans in the late 1960s. In February 1967, MacHack VI, a program written by Richard Greenblatt, an MIT undergraduate, drew one game and lost four in a United States Chess Federation tournament. In the 1990s, IBM developed and used chess to test a sophisticated new multiprocessing system, Deep Blue, using 192 custom-enhanced RS/6000 processors. Deep Blue was the first computer to defeat a world champion, Gary Kasparov. (It was later used at the 1996 Olympic Games in Atlanta, Georgia, to predict the weather.) Deep Thought, an earlier IBM chess-playing computer, by comparison, had one microprocessor and no extra chips. Deep Blue made its debut in a six-game match with PCA champion Kasparov in February 1996. Deep Blue won the first game, but Kasparov won three and drew two of the remaining games to win the match four-two. In a six-game rematch held 3–11 May 1997 in New York City, an upgraded Deep Blue was able to consider an average of two hundred million positions per second, twice its previous speed. With the match tied at one win, one loss, and three draws, Deep Blue won the decisive final game in nineteen moves.

79. The commonly held notion of building frustration is reminiscent of a model of behavior that was suggested by Konrad Lorenz. Lorenz's hydraulic model of instinctive behavior was quite influential in the 1960s. Lorenz posited that we have a

certain amount of energy available for any specific instinctual system, as illustrated by a reservoir of water. There is a neurological mechanism (innate releasing mechanism) that allows the release of some or all of that energy in the presence of the appropriate sign stimulus. See Konrad Lorenz and Agnes von Cranach, 1996, *The Natural Science of the Human Species: An Introduction to Comparative Behavioral Research: The Russian Manuscript (1944–1948)* (MIT Press, Cambridge, Mass).

80. Michael F. Smith, 1994, *Research Agenda for an Improved Novice Driver Education Program: Report to Congress* (U.S. Department of Transportation, Washington, D.C.).

81. Vest et al., Road rage.

82. Ibid.

83. Shortly after I had completed a first draft of this chapter, I received an e-mail that was so timely that it should be included:

A Smart Man

I was riding to work yesterday when I observed a female driver cut right in front of a pickup truck, causing him to have to drive onto the shoulder to avoid hitting her. This evidently angered the driver enough that he hung his arm out his window and flipped the woman off. "Man, that guy is stupid," I thought to myself.

I always smile nicely and wave in a sheepish manner whenever a woman does anything to me in traffic and here's why: I drive 48 miles each way every day to work, that's 96 miles each day. Of these, 16 miles each way is bumper-to-bumper. Most of the bumper-to-bumper is on an 8-lane highway so if you just look at the 7 lanes I am not in, that means I pass something like a new car every 40 feet per lane. That's 7 cars every 40 feet for 32 miles. That works out to be 982 cars every mile, or 31,424 cars. Even though the rest of the 32 miles is not bumper to bumper, I figure I pass at least another 4000 cars. That brings the number to something like 36,000 cars I pass every day. Statistically, half of these are driven by women, that's 18,000. In any given group of women, 1 in 28 has PMS. That's 642. According to *Cosmopolitan*, 70% describe their love life as dissatisfying or unrewarding. That's 449. According to the National Institutes of Health, 22% of all females have seriously considered suicide or homicide. That's 98. And 34% describe men as their biggest problem. That's 33. According to the National Rifle Association 5% of all females carry weapons and this number is increasing. That means that EVERY SINGLE DAY, I drive past at least one woman that has a lousy love life, thinks men are her biggest problem, has seriously considered suicide or homicide, has PMS, and is armed.

Flip one off?

I think not.

3. Beauty, Blepharoplasty, Barbie, and Miss America

1. Charles Darwin, 1871, *The Descent of Man, and Selection in Relation to Sex* (John Murray, London).

2. Reproductive value refers to a female's ability to produce offspring as well as the number of successful offspring she can produce in a lifetime.

3. David M. Buss, 1999, *Evolutionary Psychology: The New Science of the Mind* (Allyn & Bacon, Boston); D. Symons, 1979, *The Evolution of Human Sexuality* (Oxford University Press, New York); D. Symons, 1995, Beauty is in the adaptations of the beholder: the evolutionary psychology of female attractiveness, in P. R. Abramson and S. D. Pinkerton, eds., *Sexual Nature, Sexual Culture* (University of Chicago Press, Chicago), pp. 80–118.

4. Bronislaw Malinowski, 1929, Practical anthropology, *Africa: Rivista Trimestrale di Studi e Documentazione dell'Istituto Italo-Africano* 2, 22–38; Bronislaw Malinowski, 1961, *Argonauts of the Western Pacific: An Account of Native Enterprise and Adventure in the Archipelagoes of Melanesian New Guinea* (Dutton, New York). Malinowski also noted that the ideal was often approached from its negation. Deformity in mind or body, albinism, and old age put a person beyond the erotic interest (Malinowski, Practical anthropology).

5. E. Gregerson, 1983, *Sexual Practices: The Story of Human Sexuality* (Franklin Watts, New York).

6. E. Gregerson, 1994, *The World of Human Sexuality: Behaviors, Customs, and Beliefs* (Irvington Publishers, New York).

7. P. J. Brink, 1989, The fattening room among the Annang of Nigeria, *Medical Anthropology* 12 (1), 131–143.

8. J. C. Messenger, 1959, Religious acculturation among the Anang Ibibio, in W. Bascom and M. Herskovitz, eds., *Continuity and Change in African Cultures* (University of Chicago Press, Chicago), pp. 279–299.

9. L. A. Jackson, 1992, *Physical Appearance and Gender: Sociobiological and Sociocultural Perspectives* (State University of New York Press, Albany).

10. E. Berscheid and E. Walster, 1983, Physical attractiveness, in L. Berkowitz, ed., *Advances in Experimental Social Psychology* (Academic Press, New York), pp. 157–215.

11. Judith H. Langlois et al., 1991, Facial diversity and infant preferences for attractive faces, *Developmental Psychology* 27 (1), 79–84; Judith H. Langlois and Lori A. Roggman, 1990, Attractive faces are only average, *Psychological Science* 1 (2), 115–121; Judith H. Langlois et al., 1987, Infant preferences for attractive faces: rudiments of a stereotype? *Developmental Psychology* 23 (3), 363–369; Adam J. Rubenstein et al., 1999, Infant preferences for attractive faces: a cognitive explanation, *Developmental Psychology* 35 (3), 848–855; Curtis A. Samuels et al., 1994, Facial aesthetics: babies prefer attractiveness to symmetry, *Perception* 23 (7), 823–831. Attractiveness of both the black and white individuals in the photographs was independently rated by judges.

12. Michael R. Cunningham et al., 1995, "Their ideas of beauty are, on the whole, the same as ours": consistency and variability in the cross-cultural perception of female physical attractiveness, *Journal of Personality & Social Psychology* 68 (2), 261–279.

13. George Ade (1866–1944), American humorist and playwright, observed the importance of symmetry when he noted, "Her features did not know the value of

teamwork." Randy Thornhill and Karl Grammer, 1999, The body and face of woman: one ornament that signals quality? *Evolution & Human Behavior* 20 (2), 105–120.

14. S. W. Gangestad et al., 1994, Facial attractiveness, developmental stability, and fluctuating asymmetry, *Ethology and Sociobiology* 15, 73–85; Judith H. Langlois et al., 1990, Infants' differential social responses to attractive and unattractive faces, *Developmental Psychology* 26 (1), 153–159.

15. Gangestad et al., Facial attractiveness, developmental stability, and fluctuating asymmetry; K. Grammer and R. Thornhill, 1994, Human (*Homo sapiens*) facial attractiveness and sexual selection: the role of symmetry and averageness, *Journal of Comparative Psychology* (formerly *Journal of Comparative and Physiological Psychology*) 108 (3), 233–242.

16. Todd K. Shackelford and David M. Buss, 1997, Anticipation of marital dissolution as a consequence of spousal infidelity, *Journal of Social & Personal Relationships* 14 (6), 793–808.

17. There is some evidence that symmetry may not be the only basis for judgments of attractiveness. See Judith H. Langlois et al., 1994, What is average and what is not average about attractive faces? *Psychological Science* 5 (4), 214–220. Some asymmetric faces were found to have high attractiveness ratings and some symmetric faces low ratings. These results suggest that, while symmetry may be a kind of epigenetic rule in assessing attractiveness, there is still plenty of room for cultural input.

18. Following A. Fallon, 1990, Culture in the mirror: sociocultural determinants of body image, in T. F. Cash and T. Pruzinsky, eds., *Body Images: Development, Deviance, and Change* (Guilford Press, New York), pp. 80–109.

19. Tamas Bereczkei et al., 1997, Resources, attractiveness, family commitment: reproductive decisions in human mate choice, *Ethology* 103 (8), 681–699; David M. Buss, 1991, Do women have evolved mate preferences for men with resources? A reply to Smuts, *Ethology & Sociobiology* 12 (5), 401–408; S. S. Malamba et al., 1994, Risk factors for HIV-1 infection in adults in a rural Ugandan community: a case-control study, *AIDS* 8 (2), 253–257; E. O. Smith, 1987, Deception and evolutionary biology, *Cultural Anthropology* 2, 50–64; J. M. Townsend and G. D. Levy, 1990, Effects of potential partners' physical attractiveness and socioeconomic status on sexuality and partner selection, *Archives of Sexual Behavior* 19, 149–164.

20. The first commercial radio station, KDKA in Pittsburgh, signed on the air in 1920. Phyllis Stark, 1994, A history of radio broadcasting, *Billboard* 107 (1 November), 121–125.

21. Jean Kilbourne, 1994, Still killing us softly: advertising and the obsession with thinness, in Patricia Fallon and Melanie A. Katzman, eds., *Feminist Perspectives on Eating Disorders* (Guilford Press, New York), pp. 395–418; Mark Landler et al., 1991, What happened to advertising? *Business Week*, 23 September, #3232, 66–72.

22. Gerald J. Gorn and Renee Florsheim, 1985, The effects of commercials for adult products on children, *Journal of Consumer Research* 11 (4), 962–967; Gerald J. Gorn and Marvin E. Goldberg, 1987, Television and children's food habits: a big brother/sister approach, in Michael E. Manley-Casimir and Carmen Luke, eds.,

Children and Television: A Challenge for Education (Praeger, New York), pp. 34–48.

23. P. D. Fosarelli, 1984, Televison and children: a review, *Journal of Developmental and Behavioral Pediatrics* 5, 30–37; A. S. Honig, 1983, Research in review: television and young children, *Young Child* 38, 63–74; G. W. Peterson and D. F. Peters, 1983, Adolescents' construction of social reality—the impact of television and peers, *Youth and Society* 15, 67–85; M. D. Rothenberg, 1983, The role of television in shaping the attitudes of children, *Journal of the American Academy of Child Psychology* 22, 86–87; D. M. Zuckerman and B. S. Zuckerman, 1985, Television's impact on children, *Pediatrics* 75, 233–240.

24. One day, President and Mrs. Coolidge were visiting a government farm. Soon after their arrival, they were taken off on separate tours. When Mrs. Coolidge passed the chicken pens, she paused to ask the tour guide in charge if the rooster copulates more than once a day. "Dozens of times," was the reply. "Please tell that to the president," Mrs. Coolidge requested. When the president passed the pens and was told about the rooster, he asked, "Same hen every time?" "Oh no, Mr. President, a different one each time." The president nodded slowly, then said, "Tell that to Mrs. Coolidge." G. Bermant, 1976, Sexual behavior: hard times with the Coolidge effect, in M. H. Siegel and H. P. Zeigler, eds., *Psychological Research: The Inside Story* (Harper and Row, New York), pp. 76–103; Symons, *The Evolution of Human Sexuality*.

25. The whole question of gullibility is beyond the scope of this chapter, but for an interesting discussion see Albert Ellis, 1991, Suggestibility, irrational beliefs, and emotional disturbance, in John F. Schumaker, ed., *Human Suggestibility: Advances in Theory, Research, and Application* (Taylor & Francis/Routledge, Florence, Ky.), pp. 309–325.

26. The change in women's fashions was thought to be driven by changes in contemporary political context. Alfred L. Kroeber was the first anthropologist to recognize changes in women's dress styles as related to political conflict; see J. Richardson and A. L. Kroeber, 1940, Three centuries of women's dress fashions: a quantitative analysis, *Anthropological Records* 5, 111–150. Simonton investigated the effects of warfare on women's dress and concluded that dresses could be divided into two general types: the "Empire" mode was characterized by a high, wide waist with a wide skirt and décolletage; the "Hour Glass" mode was characterized by a low, narrow waist with a wide skirt and décolletage. Simonton found that during years of international war the Empire style was the most likely to dominate, while in times of international peace the Hour Glass style prevailed. For intranational war the precise opposite trend holds true. He did not offer an explanation for the observed differences but speculated that wearing short skirts during wartime was an economic issue, and the changes in the Empire style and the Hour Glass reflected greater or lesser degrees of emphasis on sexual attractiveness. See D. K. Simonton, 1977, Women's fashions and war: a quantitative comment, *Social Behavior & Personality* 5 (2), 285–288.

27. Much of this discussion of the changes in body images across time follows A. Mazur, 1986, U.S. trends in feminine beauty and overadaptation, *Journal of Sex Research* 22 (3), 281–303.

28. Ibid.

29. *The National Police Gazette: The Leading Illustrated Sporting Journal in America* was founded in 1846, and from then until 1879 it was a weekly newspaper. In the early years it was a crime tabloid, but in 1876 Richard K. Fox bought the paper and three years later changed the format to include a sporting news and theater section. It is best thought of as a precursor to today's *National Enquirer*.

30. Paul Chabas (1869–1937) was a French academic painter. He is remembered chiefly for his nude *September Morn*. It was sold to a Russian, hidden during the Russian Revolution, and rediscovered in a private collection in Paris in 1935.

31. D. E. Morrison and C. P. Holden, 1971, The burning bra: the American breast fetish and women's liberation, in P. K. Manning, ed., *Deviance and Change* (Prentice-Hall, Englewood Cliffs, N.J.), pp. 1–36.

32. Joaquin Alberto y Chavez was born on 9 February 1896 in the mountain city of Arequipa, Peru. Alberto was the first of six children born to Margarita and Max Vargas, a prominent photographer known throughout Peru for his portraits and landscapes. Like most upper-class Peruvian children, Alberto was educated in Switzerland, learning French and German and working as an apprentice in a prestigious Geneva studio. After completing his studies in Switzerland, he moved to New York, where he found employment as the official painter for the Ziegfeld Follies. In 1940, he replaced pinup artist George Petty at *Esquire Magazine*. While at Esquire Vargas dropped the "s" in his name, and an American icon, the "Varga Girl," was created. The watercolor and airbrush pinups he created for Esquire from 1940 until 1947 proved to be his most significant work. During this period the Varga Girl became a ubiquitous emblem of wartime America. She appeared on the noses of aircraft, submarines, and torpedoes, as well as in calendars, playing cards, and *Esquire Magazine*. Vargas died in California in 1982. Paul Chutkow, 1999, The real Vargas: one hundred years after his birth, Alberto Vargas is still regarded as "The King of Pin-up Art," and therein hangs an exquisite love story—and a terrible tragedy, *Cigar Aficionado* 9 (1), pp. 23–24.

33. Fallon, Culture in the mirror.

34. Ibid.

35. University of Alberta Health Centre, 2001, Health information page: food, weight and body image, *http://www.ualberta.ca/healthinfo*.

36. SBRnet, 2001, Total market consumer expenditures, *http://www.sbrnet. com/Research/Research.cfm?subRID=126*. American Society of Plastic Surgeons, 2001, 2000 average surgeon/physician fees, cosmetic surgery, *http://www. plasticsurgery.org/mediactr/average.pdf*.

37. Blaise Pascal, French scientist and polymath, noted that "had Cleopatra's nose been shorter, the whole face of the world would have been different."

38. Hence the parasol, a lightweight umbrella for protection from the sun, largely used by women, first appeared in the late seventeenth century.

39. Ginia Bellafante, 2001, The way we live: 8-12-01; tan is the new black, *New York Times Magazine*, 12 August.

40. Kathy Lee Peiss, 1998, *Hope in a Jar: The Making of America's Beauty Culture* (Metropolitan Books, New York).

41. *Lupus vulgaris* is a rare form of tuberculosis of the skin characterized by brownish tubercles that often heal slowly and leave scars.

42. In non-Western societies there may be a preference for lighter-colored skin but not necessarily youthful skin. In India, for example, signs of aging are not reviled as in the United States; witness the absence of wrinkle crèmes in the Indian market. There are, however, dozens of "Fair and Lovely" crèmes on the market.

43. H. W. Randle, 1997, Suntanning: differences in perceptions throughout history, *Mayo Clinic Proceedings* 72 (5), 461–466.

44. Data are from 1996. As an interesting aside, Americans spent $140 million on insect repellent in 1996.

45. Bellafante, The way we live: 8-12-01; tan is the new black.

46. S. Arthey and V. A. Clarke, 1995, Suntanning and sun protection: a review of the psychological literature, *Social Science & Medicine* 40 (2), 265–274.

47. M. Broadstock et al., 1992, Effects of suntan on judgements of healthiness and attractiveness by adolescents, *Journal of Applied Social Psychology* 22 (2), 157–172. Centers for Disease Control and Prevention, 2001, Facts and statistics about skin cancer, *http://www.cdc.gov/chooseyourcover/skin.htm*.

48. T. Lower et al., 1998, The prevalence and predictors of solar protection use among adolescents, *Preventive Medicine* 27 (3), 391–399; K. D. Reynolds et al., 1996, Predictors of sun exposure in adolescents in a southeastern U.S. population, *Journal of Adolescent Health* 19 (6), 409–415.

49. Broadstock et al., Effects of suntan on judgements of healthiness and attractiveness by adolescents.

50. P. H. Bloch and M. L. Richins, 1992, You look "mahvelous": the pursuit of beauty and the marketing concept, *Psychology & Marketing* 9 (1), 3–15.

51. Hazel Rawson Cades, 1927, *Any Girl Can Be Good-Looking* (D. Appleton, New York).

52. R. T. Peterson, 1987, Bulimia and anorexia in an advertising context, *Journal of Business Ethics* 6, 495–504; A. S. Tan, 1979, TV beauty ads and role expectations of adolescent female viewers, *Journalism Quarterly* 56, 283–288; E. Wilson, 1985, *Adorned in Dreams: Fashion and Modernity* (Virago Press, London).

53. R. J. Freedman, 1984, Reflections on beauty as it relates to health in adolescent females, *Women & Health* 9, 29–45; David M. Garner et al., 1980. Cultural expectations of thinness in women, *Psychological Reports* 47 (2), 483–491.

54. M. L. Richins, 1981, Social comparison and idealized images of advertising, *Journal of Consumer Research* 18, 71–83.

55. W. Feldman et al., 1988, Culture versus biology: children's attitudes toward thinness and fatness, *Pediatrics* 81 (2), 190–194; Brian E. Vaughn and Judith H. Langlois, 1983, Physical attractiveness as a correlate of peer status and social competence in preschool children, *Developmental Psychology* 19 (4), 561–567.

56. Barbie accounts for 40 percent of total sales for Mattel and 55 percent of the profits.

57. One of the keys to the success of Barbie is the variety of versions that have been marketed. Over the past forty years, there have been more than two hundred different Barbies, and some of the more notable include "Enchanted Evening"

Barbie complete with a satin strapless gown and a white "fur" stole; "Campus Sweetheart" Barbie; "Graduation" Barbie; "American Airlines Stewardess" Barbie; "Nurse" Barbie, assistant to "Dr. Ken"; "Little Red Riding Hood" Barbie; "Talking" Barbie; "Malibu" Barbie; "Sun Valley" Barbie; "NSYNC #1 Fan™" Barbie; "Sugarplum Princess" Barbie in the Nutcracker; and "Newport Beach" Barbie, to mention just a few. Ruth Handler and Jacqueline Shannon, 1994, *Dream Doll: The Ruth Handler Story* (Longmeadow Press, Stamford, Conn.).

58. L. Bannon, 1997, She reinvented Barbie; now can Jill Bardo do the same for Mattel? *Wall Street Journal*, 5 March, final edition, section A, p. 1.

59. A. M. Magro, 1997, Why Barbie is perceived as beautiful, *Perceptual & Motor Skills* 85 (1), 363–374.

60. The dismal failure of the "Happy To Be Me" doll designed and marketed by Cathy Meredig had a thicker waist, shorter neck, and shorter legs than Barbie, and she stood flat-footed. Although marketed as a healthy alternative to Barbie, the dolls had no effect on Barbie sales. (Ibid.)

61. Interestingly, "Cheerleader" Barbie has invaded college campuses across the nation. Nineteen universities (including Arizona, Arkansas, Auburn, Clemson, Florida, Georgetown, Georgia, Illinois, Miami, Michigan, Nebraska, North Carolina State, Penn State, Tennessee, Texas, Virginia, Wisconsin) have agreed to allow the dolls to wear their logos for an 8 percent royalty on the sales. Now there are plans to market "MLB Baseball" Barbies complete with a miniature Wilson glove, a Rawlings ball, and a wooden Louisville Slugger bat in the uniform of each of the Major League Baseball teams. There are plans for Barbie to play in the Women's National Basketball Association (WNBA Barbie), compete with top drivers in the National Association for Stock Car Auto Racing (NASCAR Barbie), and shoot for goals as "Women's World Cup Soccer" Barbie (WWCS Barbie). Ciro Scotti, 1999, Just what America needs: Major League Baseball Barbie, *Business Week*, 9 June, *http://www.businessweek.com/bwdaily/dnflash/june1999/nf90609d.htm*.

62. K. D. Brownell and M. A. Napolitano, 1995, Distorting reality for children: body size proportions of Barbie and Ken dolls, *International Journal of Eating Disorders* 18 (3), 295–298.

63. Magro, Why Barbie is perceived as beautiful.

64. Rebecca Brown, 1993, Barbie comes out, in Lucinda Ebersole and Richard Peabody, eds., *Mondo Barbie* (St. Martin's, New York); Richard Grayson, 1993, Twelve-step Barbie, in ibid.; Wayne E. Kline, 1993, Bar Barbie, in ibid.; Cookie Lupton, 1993, Barbie meets the scariest fatso yet, in ibid.; Belinda Subraman, 1993, The black lace panties triangle, in ibid.; Jeff Weddle, 1993, Hell's Angel Barbie, in ibid.

65. John Liggett, 1974, *The Human Face* (Constable, London).

66. Ceruse was a cosmetic that used white lead as pigment.

67. Perhaps the most notable victim of cosmetic toxicity was Maria Gunning, later the countess of Coventry, whose obstinate use of lead-based ceruse hastened her death at age twenty-seven. Liggett, *The Human Face*, p. 68.

68. T. C. Hansard, 1806, *The Parliamentary History of England from the Earliest Period to the Year 1803, from Which Last-Mentioned Epoch It Is Continued*

Downwards in the Work Entitled "Hansard's Parliamentary Debates" (T. C. Hansard, London).

69. Alan Farnham, 1996, Male vanity: you're so vain I bet you think this story's about you: American men, refusing to age gracefully, are spending $9.5 billion a year on fact-lifts, hairpieces, and makeup, *Fortune* 134 (5), 66–73.

70. Bureau of Labor Statistics, 2000, Occupational outlook handbook, 2000–01: service occupations: barbers, cosmetologists and related workers, *http://stats.bls. gov/oco/ocos169.htm*.

71. M. Merker, 1904, *Die Masai. Ethnographische Monographie Eines Ost- afrikanischen Semitenvolkes* (D. Reimer, Berlin).

72. My history of hair discussion follows Esther R. Teller, 2000, Hairdressing, *http://www.comptons.com/ceo99-cgi/article?'fastweb?getdoc+viewcomptons+ A+3537+0++'history%20of%20hair'*

73. One of the most impressively titled documents was an encyclopedic tome written by John Bulwer (1650) titled *Anthropometamorphosis: Man Transform'd; or the Artificial Changeling. Historically Presented, in the Mad Cruel Gallantry, Fool- ish Bravery, Ridiculous Beauty, Filthy Finesse, and Loathsome Lovelinesse of Most Nations, Fashioning and Altering Their Bodies from the Mould Intended by Nature. With a Vindication of the Regular Beauty and Honesty of Nature.*

74. Consumption is the name used to refer to the wasting that is associated with pulmonary tuberculosis.

75. David Kunzle, 1982, *Fashion and Fetishism: A Social History of the Corset, Tight-Lacing, and Other Forms of Body-Sculpture in the West* (Rowman and Lit- tlefield, Totowa, N.J.).

76. Fallon, Culture in the mirror.

77. Kunzle, *Fashion and Fetishism.*

78. Much of the cranial deformation discussion follows Encyclopædia Britannica, 2001, Body modification and mutilation, *http://www.britannica.com/eb/article? eu=82520&tocid=8270&query=cranial%20deformation*.

79. These people are often referred to as Ubangi after the name erroneously ap- peared in P. T. Barnum's publicity.

80. Liggett, *The Human Face*, pp. 48–49.

81. David Livingstone, 1872, *Livingstone's Africa: Perilous Adventures and Ex- tensive Discoveries in the Interior of Africa* (Hubbard Brothers, Philadelphia); David Livingstone and Charles Livingstone, 1893, *Narrative of an Expedition to the Zambesi and Its Tributaries; and of the Discovery of the Lakes Shirwa and Nyassa, 1858–1864* (Harper & Brothers, New York).

82. Laura Bohannan and Paul Bohannan, 1953, *The Tiv of Central Nigeria* (In- ternational African Institute, London).

83. Encyclopædia Britannica, 2001, Duel, *http://www.britannica.com/eb/article? eu=31895&tocid=0&query=duelling%20scars*.

84. Elizabeth Austin, 1999, Marks of mystery, *Psychology Today* 32, 46–50; Gregerson, *The World of Human Sexuality*; Kevin McAleer, 1994, *Dueling: The Cult of Honor in Fin-de-Siècle Germany* (Princeton University Press, Princeton, N.J.).

85. Darwin, *The Descent of Man, and Selection in Relation to Sex.*

86. The actual route of diffusion is unclear, whether it was by the nomadic Ainu of Japan, the Samoans, or by South American explorers. Clinton R. Sanders, 1989, *Customizing the Body: The Art and Culture of Tattooing* (Temple University Press, Philadelphia).

87. Pita Graham, 1994, *Maori Moko or Tattoo* (Bush Press, Auckland, N.Z.); Michael King and Marti Friedlander, 1972, *Moko: Maori Tattooing in the 20th Century* (Alister Taylor, Wellington); Anne Nicholas, 1994, *The Art of the New Zealand Tattoo* (Carol Pub. Group, Secaucus, N.J.); D. R. Simmons, 1986, *Ta Moko: The Art of Maori Tattoo* (Reed Methuen, Auckland, N.Z.); Iti Wairere et al., 1999, *Moko—Maori Tattoo* (Edition Stemmle, Zurich).

88. One of the most well-known tattooists was Hori Chyo, who lived in Yokohama, Japan. He was referred to as the "Shakespeare of Tattooing" by the *New York World* and was sought out by the nobility for the quality and delicacy of his tattoos.

89. Remember Mel Gibson and the battle scenes in the movie *Braveheart.*

90. Sanders, *Customizing the Body*, p. 13.

91. Gian Paolo Barbieri, 1998, *Tahiti Tattoos* (Taschen, New York); Mark Blackburn, 1999, *Tattoos from Paradise: Traditional Polynesian Patterns* (Schiffer, Atglen, Pa.); Willowdean Chatterson Handy, 1922, *Tattooing in the Marquesas* (The Honolulu Museum Press, Honolulu); Willowdean Chatterson Handy, 1965, *Forever the Land of Men: An Account of a Visit to the Marquesas Islands* (Dodd Mead, New York); Carl Marquardt, 1984, *The Tattooing of Both Sexes in Samoa* (R. McMillan, Papakura); Dirk R. V. Spennemann, 1992, *Marshallese Tattoos* (Republic of the Marshall Islands Historic Preservation Office, Majuro Atoll).

92. Hanns Ebensten, 1953, *Pierced Hearts and True Love* (D. Verschoyle, London); John Smith and Philip L. Barbour, 1986, *The Complete Works of Captain John Smith (1580–1631)* (University of North Carolina Press, Chapel Hill).

93. Steve Gilbert, 2000, *The Tattoo History Source Book* (Juno Books, New York).

94. Among the most famous were Czar Nicholas II of Russia, King George II of Greece, King Oscar II of Sweden, Kaiser Wilhelm of Germany, Don Juan of Spain, King Frederick VI of Denmark, Field Marshal George Montgomery, and most of the male members of the British royal family. Sanders, *Customizing the Body*, p. 15.

95. Leviticus 19:28.

96. While it may have been considered a deviant practice, a female tattooist working on the beaches of Atlantic City is said to have executed over ten thousand designs in a single year, 1925. Ebensten, *Pierced Hearts and True Love.*

97. William Graham Sumner, 1959, *Folkways: A Study of the Sociological Importance of Usages, Manners, Customs, Mores, and Morals* (Dover Publications, New York).

98. Liggett, *The Human Face.*

99. The probability that you will get hepatitis from a needle-stick injury is about 20 percent, while it is approximately 0.5 percent for AIDS. It takes only about 0.00004 milliliters of blood infected with hepatitis B to infect a person, while to become infected with AIDS requires about 0.1 milliliters.

100. Now that could produce ring-around-the-collar with a vengeance.

101. The maximum length was reported to be fifteen and one-half inches. Desmond Morris, 1977, *Manwatching: A Field Guide to Human Behaviour* (Cape, London).

102. Grieg's disease is a mental and physical retardation associated with a deformity of the frontal area of the skull and characterized by a low forehead, wide bridge of the nose, increased distance between the eyes, and divergent strabismus (squint). Health Encyclopedia, 2001, Grieg's disease, *http://www.lineone.net/health_ encyclopaedia/dict/pages/g/299.htm.*

103. Encyclopedia Inforplease, 2001, Gutta percha, *http://www.infoplease.com/ ce6/sci/A0822209.html.*

104. B. O. Rogers, 1971, A brief history of cosmetic surgery, *Surgical Clinics of North America* 51 (2), 265–288; B. O. Rogers, 1971, A chronologic history of cosmetic surgery, *Bulletin of the New York Academy of Medicine* 47 (3), 265–302.

105. Perhaps the best example is *The Pilgrim's Progress* by John Bunyan, first published in London in 1684. Pilgrim Christian, on his journey to the Celestial City, is confronted by a series of temptations representing some aspect of the physical world. These temptations are a distraction from his intended higher moral purpose. Christian succeeds in not succumbing to the temptations and is actually strengthened by their challenge. Ultimately he reaches his destination, the final triumph of good over evil.

106. Elizabeth Haiken, 1997, *Venus Envy: A History of Cosmetic Surgery* (Johns Hopkins University Press, Baltimore). The Dionne quintuplets were born on a farm near Callander, Ontario, Canada, on 28 May 1934. When they were four months old, the Canadian government took the girls from their parents in order to protect them from potential exploitation by American promoters. The infants were placed in a nearby hospital and were paraded before paying spectators as many as four times a day.

107. Oscar Wilde (1846–1900), British wit, poet, and dramatist, noted about women and youth, "As long as a woman can look ten years younger than her own daughter, then she is perfectly satisfied."

108. A number of famous personalities have had cosmetic surgery. Cher and Loni Anderson had breast reductions, while Mariel Hemingway had breast augmentation. Sissey Spacek and Raquel Welsh had nose jobs. Jackie Onassis, Jack Lemmon, Betty Ford, Abbie Hoffman, Burt Lancaster, and Queen Elizabeth all had face-lifts.

109. Haiken, *Venus Envy*, p. 176.

110. Sam Emerson, 1984, The Last Word on Michael, *People* (November/ December), 97.

111. For women who wear size 16 or larger.

112. Ms. Senior Sweetheart of American for women over fifty-nine, Pampered Princess or Prince up to twenty-three months, Miss Deaf Virginia, and Miss Wheelchair of Florida to name a few.

113. C. J. Leppa, 1990, Cosmetic surgery and the motivation for health and beauty, *Nursing Forum* 25 (1), 25–31. Elphaba, an emerald-green-skinned young

woman, is born into the family of a preacher and his wife in Munchkinland. Elphaba's family are not Munchkinlanders, however, and Elphaba grows up knowing more than she ever wanted to know about persecution and alienation.

114. American Academy of Cosmetic Surgery, 2001, Statistics of cosmetic procedures, *http://www.cosmeticsurgery.org/consumer/stats/2000statistics*.

115. Adipose tissue is made up of fat cells (adipocytes) in fibrocellular matrix. Fat cells are important in that they are a ready source of energy and also serve as insulation for sensitive internal organs. What we recognize as fat in most individuals is a result of an increase in the size of fat cells (hypertrophy) while fat cell numbers remain constant. In individuals who are truly obese, the adipocytes increase not only in size but also in number (hyperplasia) to accommodate the extra liquid. Hyperplastic obesity is relatively resistant to exercise and diet, while hypertrophic obesity is not resistant. Hypertrophic obesity is the more common of the two. V. J. Ablaza et al., 1998, Ultrasound assisted lipoplasty—part 1: an overview for nurses, *Plastic Surgical Nursing* 18 (1), 13–15.

116. Ring Lardner (1885–1933), American humorist and short story writer, recognized the importance of localized fatty deposits and physical appearance, especially for women, when he said, "An optimist is a girl who mistakes a bulge for a curve."

117. Leppa, Cosmetic surgery and the motivation for health and beauty.

118. Richard C. Prielipp and Robert C. Morell, 1999, Liposuction in the United States: beauty and the beast, *Anesthesia Patient Safety Foundation Newsletter* 14 (2), 18–19; Frederick M. Grazier and Rudolph H. De Jong, 2000, Fatal outcomes from liposuction: census survey of cosmetic surgeons, *Plastic and Reconstructive Surgery* 105 (1), 436–446; Arialdi M. Minino and Betty L. Smith, 2001, Death: preliminary data for 2000, *National Vital Statistics Reports* 49 (12), 14–16.

119. R. B. Rao et al., 1999, Deaths related to liposuction, *New England Journal of Medicine* 340 (19), 1471–1475.

120. John A. McCurdy, 1992, *The Complete Guide to Cosmetic Facial Surgery* (Lifetime Books, Hollywood, Fla.).

121. R. Harvey, 1998, Cosmetic surgery's ugly side, *The Toronto Star*, 13 June.

122. D. Gliddon, 1983, Through the looking glasses of physicians, dentists, and patients, *Perspectives in Biology and Medicine* 26, 451–458.

4. Fat, Diet, and Evolution

1. F. Xavier Pi-Sunyer, 1994, The fattening of America, *Journal of the American Medical Association* 272 (3), 238–239.

2. A. H. Mokdad et al., 1999, The spread of the obesity epidemic in the United States, 1991–1998, *Journal of the American Medical Association* 282 (16), 1519–1522.

3. Ibid.

4. Hillel Schwartz, 1986, *Never Satisfied: A Cultural History of Diets, Fantasies, and Fat* (Free Press, New York).

5. Graham attended Amherst Academy, but his arrogant personality alienated

the younger students, who quickly got him thrown out on a false charge of criminal assault on a woman. From Amherst he went to Rhode Island, where he had a nervous breakdown and married the daughter of a sea captain. Graham's lectures in Philadelphia and New York were very well received, but they were not without controversy. The American diet in the 1830s was centered on white bread and meat. Fruits and vegetables were thought to have little nutritional value. Graham took up the cause and touted the nutritional benefits of homemade bread made with whole grain wheat, which became known as Graham flour. He touted the "Graham system," which stressed a vegetarian diet and adherence to an early variation of "family values." In 1837, he was scheduled to speak in Boston, but his talk was canceled because of fear that butchers and bakers would burn the hall down. Butchers were angry that he told people that they ate too much meat, and bakers because he advocated homemade bread. Interestingly, the only venue that would allow Graham to speak in Boston was the Marlborough Hotel, the first temperance hotel in the United States. When the mayor of Boston warned that there were not enough police to protect the hotel, the owners barricaded the front door against the angry mob. Graham's followers commandeered the third floor and poured lime on the assembled mob. Graham was not universally accepted and was the subject of much criticism in local newspapers. Of course, he is most remembered for the crackers that bear his name. Stephen Nissenbaum and Sylvester Graham, 1980, *Sex, Diet, and Debility in Jacksonian America: Sylvester Graham and Health Reform* (Greenwood Press, Westport, Conn.).

6. Schwartz, *Never Satisified*.

7. During the 1920s and 1930s, there were coin-operated scales on almost every corner. The Peerless Scale Company was the largest manufacturer of coin-operated scales, and in 1929 Peerless was worth over $50 million (over $500 million in 2001 dollars). Since people always had a penny, even in the middle of the Depression, they could always afford to weigh themselves. In a good location, a scale could earn $50–100 a month ($500–1,000 in 2001 dollars), and even in a poor location a scale could bring in $5 a month ($50 in 2001 dollars). Scales cost $50 ($500 in 2001 dollars) but quickly paid for themselves, even under the worst conditions.

The popularity of scales reached its pinnacle in the mid-1930s, when there were over 750,000 scales all across the country. As the novelty of the scale diminished, new gimmicks were designed to revitalize interest. Some scales were designed to give a small ticket with a person's weight printed on it. Fortunes were added to the tickets, and before long pictures of movie stars were used to encourage patrons to collect tickets and complete a set. Movie stars actually paid the scale companies to feature their pictures in order to promote their names. The interest in scales decreased rapidly in the 1940s. By 1946, there were only 200,000 scales in existence. The advent of the inexpensive bathroom scale likely led to the demise of the coin-operated scale; however, vandalism and the increased costs of repairing vandalized scales also contributed to their decline.

8. F. X. Pi-Sunyer, 1993, Medical hazards of obesity, *Annals of Internal Medicine* 119 (7 part 2), 655–660.

9. High-cholesterol diets were unknown in our ancestral past since plants,

insects, and wild game have little cholesterol. See S. B. Eaton and M. Konner, 1985, Paleolithic nutrition: a consideration of its nature and current implications, *New England Journal of Medicine* 312, 283–289.

10. William Bennett and Joel Gurin, 1982, *The Dieter's Dilemma* (Basic Books, New York); Susie Orbach, 1982, *Fat Is a Feminist Issue II: A Program to Conquer Compulsive Eating* (Berkley Books, New York).

11. G. L. Maddox et al., 1968, Overweight as social deviance and disability, *Journal of Health & Social Behavior* 9 (4), 287–298; G. L. Maddox and V. Liederman, 1969, Overweight as a social disability with medical implications, *Journal of Medical Education* 44 (3), 214–220; J. H. Price et al., 1987, Family practice physicians' beliefs, attitudes, and practices regarding obesity, *American Journal of Preventive Medicine* 3 (6), 339–345.

12. P. Blumberg and L. P. Mellis, 1980, Medical students' attitudes toward the obese and the morbidly obese, *International Journal of Eating Disorders* 4, 169–175.

13. J. M. Najman et al., 1982, Patient characteristics negatively stereotyped by doctors, *Social Science & Medicine* 16 (20), 1781–1789.

14. S. L. Gortmaker et al., 1993, Social and economic consequences of overweight in adolescence and young adulthood, *New England Journal of Medicine* 329 (14), 1008–1012.

15. A. M. Wolf and G. A. Colditz, 1998, Current estimates of the economic cost of obesity in the United States, *Obesity Research* 6 (2), 97–106.

16. National Center for Health Statistics (U.S.), 1996, *The Third National Health and Nutrition Examination Survey (NHANES III, 1988–94), Reference Manuals and Reports* (U.S. Dept. of Health and Human Services, Centers for Disease Control and Prevention, National Center for Health Statistics, U.S. Government Printing Office, Washington, D.C.).

17. P. R. Thomas and Institute of Medicine (U.S.), Committee to Develop Criteria for Evaluating the Outcomes of Approaches to Prevent and Treat Obesity, 1995, *Weighing the Options: Criteria for Evaluating Weight-Management Programs* (National Academy Press, Washington, D.C.).

18. "Minority populations" refers to African-Americans; Hispanic-Americans (including Mexican-Americans, Puerto Ricans, Cuban-Americans, Central Americans, and others); Asian and Pacific Islanders (including individuals with origins in East or Southeast Asia, the Indian subcontinent, or the Pacific Islands); American Indians and Alaska Natives (including persons living both on and off reservations, among the more than five hundred federally recognized Indian tribes as well as the Eskimos, Alaskan Indians, and Aleuts); and Native Hawaiians. S. K. Kumanyika, 1993, Special issues regarding obesity in minority populations, *Annals of Internal Medicine* 119 (7 part 2), 650–654.

19. Thomas and Institute of Medicine (U.S.), *Weighing the Options*.

20. Caroline Davis and Etta Saltos, 1999, Dietary recommendations and how they have changed over time, in Elizabeth Frazão, ed., *America's Eating Habits: Changes and Consequences*, Agriculture Information Bulletin no. 750 (Food and Rural Economics Division, Economic Research Services, USDA, Washington, D.C.).

21. Hamilton Cravens, 1990, Establishing the science of nutrition at the USDA: Ellen Swallow Richards and her allies, *Agricultural History* 64 (2), 122–133.

22. Data adapted from National Livestock and Meat Board and the MRCA Information Services.

23. S. B. Eaton et al., 1988, *The Paleolithic Prescription: A Program of Diet and Exercise and a Design for Living* (Harper & Row, New York).

24. USDA data points are taken as the average of the specified range in the Food Guide Pyramid.

25. Physicians Committee for Responsible Medicine, 2000. Doctors announce final victory in dietary guidelines lawsuit, Washington, D.C.

26. The subtitle to this section is taken loosely from the movie *The French Connection*, which portrays the seizure of the largest shipment of heroin into the United States up until the early 1960s. The French connection was actually Jacques Angelvin, an emcee on France's most popular television show at the time, *Paris-Club*, who allowed Sicilian loan sharks to use his 1960 Buick Invicta as the delivery vehicle. In return, one would presume that the Sicilians forgave Angelvin's debt, or at least a portion of it. The car arrived in New York on 10 January 1962 aboard the SS *United States*, and, after it was unloaded from the ship, it was driven to a garage in the Bronx and the heroin was removed. Now what does this have to do with this subtitle? The estimated street value of the ninety-seven pounds of heroin seized by New York City narcotics detectives Eddy Egan and Sonny Grosso was over $2 million (@ $50/gram), which is trivial when compared to over $16 million per day that Americans spent on fast food at about the same time.

27. Data are difficult to obtain on the precise number of fast-food outlets in hospitals. A survey of the major fast-food corporations netted data from Chick-fil-A, based in Atlanta, Georgia. Chick-fil-A has over 850 stores, and 17 of those are located in hospitals: Atlanta Medical Center, Atlanta, Georgia; Brookwood Medical Center, Birmingham, Alabama; Central Carolina Hospital, Sanford, North Carolina; Covenant Medical Center, Lubbock, Texas; DeKalb Medical Center, Decatur, Georgia; Gadsden Hospital, Gadsden, Alabama; Greenville Memorial Hospital, Greenville, South Carolina; Mary Washington Hospital, Fredericksburg, Virginia; Medical Center of Columbus, Columbus, Georgia; Piedmont Hospital, Atlanta, Georgia; Roanoke Memorial Hospital, Roanoke, Virginia; Shands Hospital, Gainesville, Florida; St. Joseph's Hospital, Atlanta, Georgia; St. Michael's Health System, Texarkana, Texas; Tulane Medical Center, New Orleans, Louisiana; University Hospital—Augusta, Georgia; University of Arkansas Medical Science, Little Rock, Arkansas.

28. Richard L. Papiernik, 2001, Sales expected to hit $399B, *Nation's Restaurant News* 38 (1), 1–3.

29. Office of Management and Budget, 2001, *Budget of the United States Government*, 2001 (Washington, D.C.).

30. Eric Schlosser, 2001, *Fast Food Nation: The Dark Side of the All-American Meal* (Houghton Mifflin, Boston).

31. Marketing to children has a heightened importance on Madison Avenue. *Kid Trends, Targeting Teens, KidTrends 2001 Report, Marketing to Tweens 2001 Report,*

Latino Kidtrends 2001 Report, Kids' Snacks & Prepared Food 2001 Report, and the *Under Five Crowd 2001 Report*, a "Syndicated Study of Pre-Schoolers and Mothers," are all prepared especially for those marketing to children. These reports are available from Children's Market Services, Inc., for a mere $2,595 each. See Children's Market Services, Inc., 2001, Strategic planners in the youth market, *http://www.kidtrends.com*. Today, the advertising industry spends over $2 billion annually specifically on ads directed to kids. Children also influence an increasingly large percentage of parents' purchases. In 1984, $50 billion; 1997, $188 billion; and 2000, $290 billion. See James U. McNeal, 1998, Tapping the three kids' markets, *American Demographics* 20 (4), 36–42; James U. McNeal, 1999, *The Kids Market: Myths and Realities* (Paramount Market, Ithaca, N.Y.).

32. Schlosser, *Fast Food Nation*; Rod Taylor, 1997, The Beanie factor, *Brandweek* 38 (24), 22–27.

33. Michael F. Jacobson, 1999, *Liquid Candy: How Soft Drinks Are Harming Americans' Health* (Center for Science in the Public Interest, Washington, D.C.).

34. USDA, 2000, *Dietary Guidelines for Americans*, Home and Garden Report 232 (United States Department of Agriculture, Washington, D.C.).

35. There are at least forty-six varieties of potatoes grown in the United States, including such all-time favorites as the Russet Burbank, Red Pontiac, Red Warba, Irish Cobbler "BC," Norgold Russet "BC," Norland, Atlantic, White Rose, Superior, Centennial Russet, Keswick "NB 1," Green Mountain, Burbank, Pontiac, Warba, Irish Cobbler, and Dark Red Norland. Potato Association of America, 2001, Potato varieties, *http://www.css.orst.edu/classes/CSS322/potato.htm*.

36. Sucrose is the source for the accumulation of reducing sugar. Reducing sugars come from the breakdown of starch to sucrose. If one can minimize the amount of sucrose, then it is possible to control the darkening process. Unfortunately, this is easier said than done. Many environmental factors affect the sugar content of potatoes, including the conditions in the field, the variety of potato, the storage conditions after harvest, as well as the duration of storage. So the proper "curing" of potatoes becomes a real concern. If potatoes become "senescent" during storage, they will have an unacceptable sugar content. Russet Burbank potatoes are not stored below 45° F for several weeks. Particulars of the aging process are well-guarded secrets of the potato-growing community.

37. Malcolm Gladwell, 2001, The trouble with fries; fast food is killing us. Can it be fixed? *The New Yorker* 77 (3), 52; Nora Olsen, 2000, Varietal differences bring fry color challenge, *Potato Grower* 29 (12), 6–8.

38. The secret of the fries is in the temperature of the cooking oil. Well, almost. When cold raw "fries to be" are placed into the preheated cooking oil, the temperature of the oil drops, and then, as the potatoes cook, the temperature rises. Given that the initial batch of "fries to be" is a constant size, the cooking vat a constant volume, and the chemical composition of the cooking oil unchanged, it is a matter of figuring out how many degrees the cooking oil must rise to have perfect fries. And the envelope please . . . 3° F.

39. Schlosser, *Fast Food Nation*.

40. Formula 47 refers to the 47¢ "All-American meal": a 15¢ hamburger, 12¢ fries, and a 20¢ shake. Gladwell, The trouble with fries; fast food is killing us.

41. Fats are categorized based on whether or not the chemical bonds between the carbon atoms of the fat molecules contain all the hydrogen atoms they can possibly hold. Saturated fats are those that have a complete complement of hydrogen atoms. Unsaturated fats have the capacity to accept more hydrogen atoms. Unsaturated fats can be converted to saturated fats by the addition of hydrogen atoms in a process called hydrogenization. The resulting product is a saturated fat called trans fatty acid.

42. Associated Press, 2001, Indian beef protesters raid McDonald's, Associated Press, 6 May; Martin Hickman, 2001, McDonalds expresses "regret" in meaty fries row, *Reuters Business Report*, 24 May; Reuters Business Report, 2001, McDonald's sued by another group over beefy fries, *Reuters Business Report*, 12 June; United Press International, 2001, McDonald's secret sauce secret no more, 13 August.

43. Derek A. Denton, 1982, *The Hunger for Salt: An Anthropological, Physiological, and Medical Analysis* (Springer-Verlag, Berlin).

44. Jay Schulkin et al., 1984, Salt hunger in the rhesus monkey, *Behavioral Neuroscience* 98 (4), 753–756. Although not without some critics. Teresa M. McMurray and Charles T. Snowdon, 1977, Sodium preferences and responses to sodium deficiency in rhesus monkeys, *Physiological Psychology* 5 (4), 477–482.

45. G. K. Beauchamp et al., 1990, Experimental sodium depletion and salt taste in normal human volunteers, *American Journal of Clinical Nutrition* 51 (5), 881–889; M. Bertino et al., 1981, Taste perception in three individuals on a low sodium diet, *Appetite* 2 (1), 67–73; R. A. McCance, 1936, Medical problems in mineral metabolism III: experimental human salt deficiency, *Lancet* 230, 823–830; R. A. McCance, 1938, The effect of salt deficiency in man on the volume of the extracellular fluid and on the composition of sweat, saliva, gastric juice and cerebrospinal fluid, *Journal of Physiology* 92, 208–218.

46. J. Schulkin, 1986, The evolution and expression of salt appetite, in G. de Caro et al., eds., *The Physiology of Thirst and Appetite* (Plenum, New York).

47. Jay Schulkin, 1991, *Sodium Hunger: The Search for a Salty Taste* (Cambridge University Press, Cambridge).

48. Jesus, speaking to his disciples, tells them they are the salt of the earth. Theologians have interpreted this to refer to the preservative properties of salt. Where there is no refrigeration, it is not uncommon to cure meat or fish by treating it with salt. Jesus is referring to humanity being in a state of decay, and his followers can save it (Matthew 5:13).

49. David Bloch, 1999, Salt and gold, *Journal of Salt History* 7, 9.

50. J. George Fodor et al., 1999, Recommendations on dietary salt, *Canadian Medical Association Journal* 160 (9 supplement), S29–S34.

51. Richard S. Cooper et al., 1999, The puzzle of hypertension in African-Americans, *Scientific American* 280 (2), 56–63; Richard S. Cooper and Charles N. Rotimi, 1994, Hypertension in populations of West African origin: is there a genetic predisposition? *Journal of Hypertension* 12 (3), 215–227; Richard S. Cooper et al.,

1997, Hypertension prevalence in seven populations of African origin, *American Journal of Public Health* 87 (2), 160–168.

52. Graham A. MacGregor and Peter S. Sever, 1996, Salt—overwhelming evidence but still no action: can a consensus be reached with the food industry? *British Medical Journal* 312 (7041), 1287–1289; D. S. Thelle, 1996, Salt and blood pressure revisited, *British Medical Journal* 312 (7041), 1240–1241.

53. M. H. Alderman et al., 1998, Dietary sodium intake and mortality: the national health and nutrition examination survey (NHANES I), *Lancet* 351 (9105), 781–785.

54. R. L. Hanneman, 1996, Intersalt: hypertension rise with age revisited, *Journal of the American Medical Association* 312 (7041), 1283–1284; discussion, 1284–1287; T. A. Kotchen and R. M. Krauss, 1996, Dietary sodium and blood pressure, *Journal of the American Medical Association* 276 (18), 1468; discussion, 1469–1470; G. A. MacGregor and F. P. Cappuccio, 1996, Dietary sodium and blood pressure, *Journal of the American Medical Association* 276 (18), 1468–1469; discussion, 1469–1470; F. H. Messerli and R. E. Schmieder, 1996, Dietary sodium and blood pressure, *Journal of the American Medical Association* 276 (18), 1469; discussion, 1469–1470; J. Stamler, 1996, Dietary sodium and blood pressure, *Journal of the American Medical Association* 276 (18), 1467; discussion, 1469–1470; P. K. Whelton et al., 1996, Dietary sodium and blood pressure, *Journal of the American Medical Association* 276 (18), 1467–1468; discussion, 1469–1470.

55. Bile is a greenish-yellow secretion produced in the liver and passed to the gallbladder for concentration, storage, or transport into the small intestine. Bile is composed of bile acids and salts, cholesterol, pigments, water, and electrolyte chemicals that keep the total solution slightly acidic (with a pH of between five and six). Bile is continually secreted from the liver into the bile duct and gallbladder; once in the gallbladder it is usually concentrated to about five times—and sometimes as high as eighteen times—the strength of the original secretion. The amount of bile secreted into the intestine is controlled by the hormones secretin, gastrin, and cholecystokinin, and by the vagus nerve. About 250 to 1,000 milliliters of bile (before concentration) are produced by the liver daily. Encyclopædia Britannica, 2001, Bile, *http://www.britannica.com/eb/article?eu=81340&tocid=0&query=bile*.

56. E. N. Marieb, 1995, *Human Anatomy and Physiology* (Benjamin/Cummings, Redwood City, Calif.).

57. Actually, it is not white but cream-colored, deep yellow, or pink.

58. Much of the preceding discussion is based on Caroline M. Pond, 1998, *The Fats of Life* (Cambridge University Press, Cambridge).

59. Pond has shown a positive correlation between body size and number of adipocytes, and regression analyses have demonstrated that larger species have fewer adipocytes in proportion to their body size than do smaller species. Pond, *The Fats of Life*.

60. Crab-eating macaques (*Macaca fascicularis*) are small-bodied, Old World monkeys that are indigenous to Southeast Asia.

61. Peter J. Brown and Melvin J. Konner, 1987, An anthropological perspective on obesity, *Annals of the New York Academy of Sciences* 499, 29–46.

62. S. M. Bernes and A. M. Kaplan, 1994, Evolution of neonatal seizures, *Pediatric Clinics of North America* 41 (5), 1069; J. R. Kaplan et al., 1997, Assessing the observed relationship between low cholesterol and violence-related mortality: implications for suicide risk, *Annals of the New York Academy of Sciences* 836, 57–80; M. F. Muldoon et al., 1992, Effects of a low-fat diet on brain serotonergic responsivity in cynomolgus monkeys, *Biological Psychiatry* 31 (7), 739–742; M. E. Berman et al., 1997, The serotonin hypothesis of aggression revisited, *Clinical Psychology Review* 17 (6), 651–665; L. Buydens-Branchey et al., 2000, Low HDL cholesterol, aggression and altered central serotonergic activity, *Psychiatry Research* 93 (2), 93–102; D. H. Edwards and E. A. Kravitz, 1997, Serotonin, social status and aggression, *Current Opinion in Neurobiology* 7 (6), 812–819; C. F. Ferris et al., 1999, Serotonin regulation of aggressive behavior in male golden hamsters (*Mesocricetus auratus*), *Behavioral Neuroscience* 113 (4), 804–815; T. R. Insel and J. T. Winslow, 1998, Serotonin and neuropeptides in affiliative behaviors, *Biological Psychiatry* 44 (3), 207–219.

63. Marieb, *Human Anatomy and Physiology*.

64. For most mammalian females, there is only a small amount of adipose tissue associated with the breasts, and so among primates and other mammals humans are unique. Pond, *The Fats of Life*.

65. Stephen M. Bailey, 1982, Absolute and relative sex differences in body composition, in Roberta L. Hall, ed., *Sexual Dimorphism in Homo sapiens: A Question of Size* (Praeger, New York), pp. 363–390.

66. L. A. Campfield et al., 1996, Human eating: evidence for a physiological basis using a modified paradigm, *Neuroscience & Biobehavioral Reviews* 20 (1), 133–137.

67. Brown fat is plentiful in hibernating animals, human infants, and cold-adapted animals, and its color is due to its high mitochondrial content. Brown fat cells can produce their own heat through hydrolyzing their ATP. Such a short circuit produces heat in the same way as mixing a strong acid and base together. Arndt von Hippel, 1994, *Human Evolutionary Biology: Human Anatomy and Physiology from an Evolutionary Perspective* (Stone Age Press, Anchorage).

68. Brown and Konner, An anthropological perspective on obesity.

69. Ibid.

70. Marieb, *Human Anatomy and Physiology*.

71. Dosage was 5 mg/kg of body weight.

72. J. L. Halaas et al., 1995, Weight-reducing effects of the plasma protein encoded by the obese gene, *Science* 269 (5223), 543–546.

73. P. Trayhurn et al., 1999, Leptin: fundamental aspects, *International Journal of Obesity & Related Metabolic Disorders* 23 (supplement 1), 22–28.

74. C. J. Hukshorn et al., 2000, Weekly subcutaneous pegylated recombinant native human leptin (peg-ob) administration in obese men, *Journal of Clinical Endocrinology & Metabolism* 85 (11), 4003–4009; E. Jequier and L. Tappy, 1999, Regulation of body weight in humans, *Physiological Reviews* 79 (2), 451–480; F. Lonnqvist et al., 1999, Leptin and its potential role in human obesity, *Journal of Internal Medicine* 245 (6), 643–652.

75. N. V. Dhurandhar et al., 2000, Increased adiposity in animals due to a human virus, *International Journal of Obesity & Related Metabolic Disorders* 24 (8), 989–996.

76. British Broadcasting Corporation, 2001, Obesity gene pinpointed, *http://news.bbc.co.uk/hi/english/health/newsid_1484000/484659.stm.*

77. S. M. Ogletree et al., 1990, Female attractiveness and eating disorders: do children's television commercials play a role? *Sex Roles* 22, 791–797.

78. The idea of genetic suicide relates directly to questions of fitness in the evolutionary sense.

79. J. J. Brumberg, 1988, *Fasting Girls: The Emergence of Anorexia Nervosa as a Modern Disease* (Harvard University Press, Cambridge, Mass.).

80. D. L. Stephens et al., 1994, The beauty myth and female consumers: the controversial role of advertising, *The Journal of Consumer Affairs* 28, 137–153.

81. T. F. Cash et al., 1986, Body image survey report: the great American shape up, *Psychology Today* 20, 30–37.

82. T. F. Cash and K. L. Hicks, 1990, Being fat versus thinking fat: relationships with body image, eating behaviors and well-being, *Cognitive Therapy and Research* 14, 327–341.

83. W. DeJong and R. E. Kleck, 1986, The social psychological effects of overweight, in P. Herman et al., eds., *Physical Appearance, Stigma, and Social Behavior: The Ontario Symposium*, vol. 3. (Lawrence Erlbaum, Hillsdale, N.J.), pp. 65–87.

84. M. Schroeder, 1991, The diet business is getting a lot skinnier, *Business Week*, 24 June, #3219, pp. 132–134.

85. SBRnet, 2001, Total market consumer expenditures, *http://www.sbrnet.com/Research/Research.cfm?subRID=126.*

86. Kalatu Kamara et al., 1998, High-fat diets and stress responsivity, *Physiology & Behavior* 64 (1), 1–6.

87. D. F. Williamson et al., 1992, Weight loss attempts in adults: goals, duration, and rate of weight loss, *American Journal of Public Health* 82 (9), 1251–1257.

88. R. B. Friedman, 1986, Fad diets: evaluation of five common types, *Postgraduate Medicine* 79 (1), 249–255.

89. Thomas and Institute of Medicine (U.S.), *Weighing the Options.*

90. F. Grodstein et al., 1996, Three-year follow-up of participants in a commercial weight loss program: can you keep it off? *Archives of Internal Medicine* 156 (12), 1302–1306.

91. Melvin Konner, 1982, *The Tangled Wing: Biological Constraints on the Human Spirit* (Holt Rinehart and Winston, New York).

92. Exercise levels from approximately 700 to 2,000 kcal of energy expenditure per week were found to be beneficial.

93. S. N. Blair et al., 1995, Changes in physical fitness and all-cause mortality: a prospective study of healthy and unhealthy men, *Journal of the American Medical Association* 273 (14), 1093–1098; I. M. Lee et al., 1995, Exercise intensity and longevity in men: the Harvard alumni health study, *Journal of the American Medical Association* 273 (15), 1179–1184.

94. Marieb, *Human Anatomy and Physiology.*

95. Only 5–10 percent of anorexics are male. H. M. Prescott, 1993, Anorexia nervosa, in K. F. Kiple, ed., *The Cambridge World History of Human Disease* (Cambridge University Press, New York), pp. 577–582.

96. Ibid.

97. D. B. Herzog and P. M. Copeland, 1985, Eating disorders, *New England Journal of Medicine* 313 (5), 295–303.

98. Kim Chernin, 1985, *The Hungry Self: Women, Eating and Identity* (Times Books, New York); Susie Orbach, 1986, *Hunger Strike: An Anorectic's Struggle as a Metaphor for Our Age* (Norton, New York).

99. H. C. Steinhausen et al., 1991, Follow-up studies of anorexia nervosa: a review of four decades of outcome research, *Psychological Medicine* 21 (2), 447–454.

100. Speech by Agriculture Secretary Dan Glickman, 1 July 1997, in San Diego, California, at the annual meeting of Second Harvest, the largest domestic charitable relief organization in the United States.

5. Depression, Antidepressants, and Evolution

1. Hara Estroff Marano, 1999, Depression: beyond serotonin, *Psychology Today*, 32 (1 April), 30–36, 72.

2. Henry E. Sigerist, 1987, *A History of Medicine: Early Greek, Hindu, and Persian Medicine* (Oxford University Press, New York).

3. Psalms 23:4.

4. Sad Sack was originally drawn for *Yank*, the magazine published for soldiers during World War II, and was used in recruitment efforts at the end of the war. "Don't be a Sad Sack, re-enlist in the Regular Army."

5. American Psychiatric Association, 1994, *Diagnostic and Statistical Manual of Mental Disorders, DSMIV* (American Psychiatric Association, Washington, D.C.).

6. William Shakespeare and George MacDonald, 1885, *The Tragedie of Hamlet, Prince of Denmarke; a Study with the Text of the Folio of 1623* (Longmans Green and Co., London).

7. Robert M. Sapolsky, 1994, *Why Zebras Don't Get Ulcers: A Guide to Stress, Stress Related Diseases, and Coping* (W. H. Freeman, New York).

8. Eric R. Kandel et al., 1991, *Principles of Neural Science* (Appleton & Lange, Norwalk, Conn.).

9. S. E. Hyman, 2000, The genetics of mental illness: implications for practice, *Bulletin of the World Health Organization* 78 (4), 455–463; K. S. Kendler and C. A. Prescott, 1999, A population-based twin study of lifetime major depression in men and women, *Archives of General Psychiatry* 56 (1), 39–44; M. J. Owen and M. J. Mullan, 1990, Molecular genetic studies of manic-depression and schizophrenia, *Trends in Neurosciences* 13 (1), 29–31; A. J. Rothschild, 1988, Biology of depression, *Medical Clinics of North America* 72 (4), 765–790; D. R. Wilson, 1998, Evolutionary epidemiology and manic depression, *British Journal of Medical Psychology* 71 (part 4), 375–395.

10. N. C. Andreasen and G. Winokur, 1979, Secondary depression: familial,

clinical, and research perspectives, *American Journal of Psychiatry* 136 (1), 62–66; Rothschild, Biology of depression.

11. Anthony Stevens and John Price, 1996, *Evolutionary Psychiatry: A New Beginning* (Routledge, London).

12. If novice observers can readily determine winners and losers in these contests, imagine how powerful the signals are to the participants.

13. The classic "agony of defeat" came on 21 March 1970, when Yugoslavian Vinko Bogataj careened off a ramp during the International Ski Flying Championship in Oberstdorf, West Germany. Bogataj not only survived the hideous spill but went on to become a celebrity thanks to ABC's *Wide World of Sports* coordinating producer Dennis Lewin, who inserted the segment into the show's opening credits. Lewin recalls the United States ski-jumping team being none too pleased with his use of Vinko's tumble: "They thought we were giving the sport a bad name." One of the classic examples of "thrill of victory" occurred at the 1976 Olympics when Nadia Comaneci made history, becoming the first gymnast to ever score a perfect ten. At those games she received seven perfect tens, three gold medals, one silver, and one bronze.

14. Randolph M. Nesse and George C. Williams, 1995, *Why We Get Sick: The New Science of Darwinian Medicine* (Times Books, New York).

15. Concorde fallacy is a concept that derives its name from the supersonic transport built jointly by aircraft manufacturers in France and Great Britain and put into regular service in 1976. In 1982, Air France drastically reduced its participation in the project due to its excessive cost. The British, however, continued to use the Concorde until a series of recent mishaps called the safety of the aircraft into question. The fallacy involved comes from microeconomics, sometimes called the problem of sunken costs: decisions to continue investing in a project because of past investment, not future payoff. R. Dawkins and T. R. Carlisle, 1976, Parental investment, mate desertion and a fallacy, *Nature* 2262, 131–133.

16. Robert Desjarlais et al., 1995, *World Mental Health: Problems and Priorities in Low-Income Countries* (Oxford University Press, Oxford).

17. R. M. Hirschfeld et al., 1997, The national depressive and manic-depressive association consensus statement on the undertreatment of depression, *Journal of the American Medical Association* 277 (4), 333–340.

18. M. Olfson and G. L. Klerman, 1992, The treatment of depression: prescribing practices of primary care physicians and psychiatrists, *Journal of Family Practice* 35 (6), 627–635.

19. Desjarlais et al., *World Mental Health*.

20. Cross-National Collaborative Group, 1992, The changing rate of major depression: cross-national comparisons, *Journal of the American Medical Association* 268 (21), 3098–3105.

21. G. E. Simon et al., 1995, Is the lifetime risk of depression actually increasing? *Journal of Clinical Epidemiology* 48 (9), 1109–1118.

22. Edward Shorter, 1997, *A History of Psychiatry: From the Era of the Asylum to the Age of Prozac* (John Wiley & Sons, New York).

23. S. M. Schappert and C. Nelson, 1999, National ambulatory medical care survey, 1995–96: summary, *Vital Health Statatistics* 13 (142), i–vi, 1–122.

24. The place of psychoactive drugs in the arts and literature is based on Marc Kusinitz, 1987, *Drugs & the Arts* (Chelsea House Publishers, New York).

25. R. Kuhn, 1989, The discovery of modern antidepressants, *Psychiatric Journal of the University of Ottawa* 14 (1), 249–252.

26. S. L. Dubovsky, 1994, Beyond the serotonin reuptake inhibitors: rationales for the development of new serotonergic agents, *Journal of Clinical Psychiatry* 55, 34–44.

27. Remember that a human hair is about 125 micrometers thick, and one nanometer is 0.001 micrometers.

28. The discussion of the history of the development of antidepressants is taken from Peter D. Kramer, 1993, *Listening to Prozac: A Psychiatrist Explores Antidepressant Drugs and the Remaking of the Self* (Viking, New York).

29. There appears to be a dose-dependent relationship in some of the adverse side effects for Paxil. John Morgan Jones, 1997, *Physicians' Desk Reference to Pharmaceutical Specialties and Biologicals: Five Sections* (Medical Economics, Rutherford, N.J.).

30. Dubovsky, Beyond the serotonin reuptake inhibitors.

31. Ibid.

32. Not only are antidepressants useful for humans, in January 1999 the FDA approved the use of antidepressants in animals. Veterinarians have used psychoactive drugs on animals for years. One of the most famous cases was Gus, the polar bear at New York's Central Park Zoo, who was prescribed Prozac after a $25,000 psychiatric consultation diagnosed the seven-hundred-pound bear with boredom neurosis. Clomicalm and Anipryl have been approved as behavior modification drugs for animals. Clomicalm is actually clomipramine hydrochloride, a tricyclic antidepressant, and is used to treat separation anxiety in dogs. It is the same as Anafranil, marketed for humans as a tricyclic antidepressant. Anipryl (selegiline hydrochloride, L-deprenyl hydrochloride) is a MAO inhibitor and is used to treat cognitive dysfunction syndrome in dogs. It is marketed as Eldepryl for human Parkinson's disease treatment. In addition to these drugs, Valium (diazepam) and Elavil (amitriptyline) are used in veterinary practice. For additional information, see Arianna Huffington, 1999, Barking back at Prozac, *Washington Times*, 20 January, Final, A, p. 15; Lauran Neergaard, 1999, FDA Approves anti-depressant for dogs, *The Columbian*, 5 January 1999, world/national news.

33. Robert Pear, 2001, Spending on prescription drugs increases by almost 19 percent, *New York Times*, 8 May, final edition, section A, p. 1; Marianne Szegedy-Maszak, 2001, The career of a celebrity pill, *U.S. News & World Report* 131 (5), p. 38.

34. National Center for Health Statistics, 2001, *National Ambulatory Medical Care Survey, 1999: Summary* (U.S. Public Health Service, Centers for Disease Control and Prevention, Washington, D.C.).

35. M. M. Katz et al., 1988, On the expression of psychosis in different cultures:

schizophrenia in an Indian and in a Nigerian community, *Culture, Medicine & Psychiatry* 12 (3), 331–355; A. J. Marsella, 1988, Cross-cultural research on severe mental disorders: issues and findings, *Acta Psychiatrica Scandinavica*, Supplementum 344, 7–22.

36. M. Olfson and G. L. Klerman, 1993, Trends in the prescription of antidepressants by office-based psychiatrists, *American Journal of Psychiatry* 150 (4), 571–577; M. Olfson et al., 1998, Antidepressant prescribing practices of outpatient psychiatrists, *Archives of General Psychiatry* 55 (4), 310–316.

37. J. L. Rushton et al., 2000, Pediatrician and family physician prescription of selective serotonin reuptake inhibitors, *Pediatrics* 105 (6), E82; J. L. Rushton and J. T. Whitmire, 2001, Pediatric stimulant and selective serotonin reuptake inhibitor prescription trends: 1992 to 1998, *Archives of Pediatrics & Adolescent Medicine* 155 (5), 560–565.

38. D. J. Bramble, 1995, Antidepressant prescription by British child psychiatrists: practice and safety issues, *Journal of the American Academy of Child & Adolescent Psychiatry* 34 (3), 327–331.

39. R. L. Fisher and S. Fisher, 1996, Antidepressants for children: is scientific support necessary? *Journal of Nervous & Mental Disease* 184 (2), 99–102. Efficacy refers to optimum response to a treatment under the best possible treatment conditions. Factors that can mitigate against this may include noncompliance, comorbidity, infrequent contact with clinician, unusual life stresses or changes, and poor social support. R. J. Goldberg, 1997, Antidepressant use in the elderly: current status of nefazodone, venlafaxine and moclobemide, *Drugs & Aging* 11 (2), 119–131.

40. N. D. Ryan, 1992, The pharmacologic treatment of child and adolescent depression, *Psychiatric Clinics of North America* 15 (1), 29–40.

41. D. J. Goldstein, 1995, Effects of third trimester fluoxetine exposure on the newborn, *Journal of Clinical Psychopharmacology* 15 (6), 417–420; D. J. Goldstein et al., 1997, Effects of first-trimester fluoxetine exposure on the newborn, *Obstetrics & Gynecology* 89 (5 part 1), 713–718; D. J. Goldstein et al., 1997, Birth outcomes in pregnant women taking fluoxetine, *New England Journal of Medicine* 336 (12), 872–873; discussion, 873; G. Koren et al., 1998, Outcome of children exposed in utero to fluoxetine: a critical review, *Depression & Anxiety* 8 (supplement 1), 27–31; M. J. Mhanna et al., 1997, Potential fluoxetine chloride (Prozac) toxicity in a newborn, *Pediatrics* 100 (1), 158–159; K. L. Wisner et al., 1999, Pharmacologic treatment of depression during pregnancy, *Journal of the American Medical Association* 282 (13), 1264–1269.

42. E. S. Paykel et al., 1997, The defeat depression campaign: psychiatry in the public arena, *American Journal of Psychiatry* 154 (6 supplement), 59–65.

43. R. P. Greenberg et al., 1994, A meta-analysis of fluoxetine outcome in the treatment of depression, *Journal of Nervous & Mental Disease* 182 (10), 547–551; D. C. Manolis, 1995, The perils of Prozac, *Minnesota Medicine* 78 (1), 19–23; Shorter, *A History of Psychiatry*.

44. L. J. Brandes, 1992, Depression, antidepressant medication, and cancer, *American Journal of Epidemiology* 136 (11), 1414–1417; L. J. Brandes and

M. Cheang, 1995, Letter regarding "response to antidepressants and cancer: cause for concern?" *Journal of Clinical Psychopharmacology* 15 (1), 84–85.

45. A recent study by Weiss and his colleagues of antidepressant use and subsequent tumor growth in a cohort of 1,467 patients from 1988 to 1994 failed to find an increased risk for recurrent or second primary tumors among the patients with cancer. Two limitations of the study are (1) its failure to specify the type of antidepressant prescribed and (2) its involving patients with only colon and breast cancer. S. R. Weiss et al., 1998, Cancer recurrences and secondary primary cancers after use of antihistamines or antidepressants, *Clinical Pharmacology & Therapeutics*, 63 (5), 594–599.

46. S. Kapur et al., 1992, Antidepressant medications and the relative risk of suicide attempt and suicide, *Journal of the American Medical Association* 268 (24), 3441–3445.

47. These are data for the United Kingdom until 1991. S. A. Montgomery, 1997, Suicide and antidepressants, *Annals of the New York Academy of Sciences* 836, 329–338.

48. Jones, *Physicians' Desk Reference to Pharmaceutical Specialties and Biologicals*.

49. Economic aspects of depression are reviewed in Hirschfeld et al., The national depressive and manic-depressive association consensus statement on the undertreatment of depression.

50. D. Eccleston, 1993, The economic evaluation of antidepressant drug-therapy, *British Journal of Psychiatry* 163, 5–6.

51. B. A. Majeroni and A. Hess, 1998, The pharmacologic treatment of depression, *Journal of the American Board of Family Practice* 11 (2), 127–139.

52. J. K. Aronson and M. Hardman, 1992, ABC of monitoring drug therapy: patient compliance, *British Medical Journal* 305 (6860), 1009–1011; J. K. Aronson and M. Hardman, 1992, ABC of monitoring drug therapy: why monitor drug therapy? *British Medical Journal* 305 (6859), 947–948.

53. R. Lane, 1993, Managing depression in general practice, *British Medical Journal* 306 (6872), 272–273.

54. C. Thompson et al., 2000, Compliance with antidepressant medication in the treatment of major depressive disorder in primary care: a randomized comparison of fluoxetine and a tricyclic antidepressant, *American Journal of Psychiatry* 157 (3), 338–343.

55. J. Sorensen and P. Kind, 1995, Modelling cost-effectiveness issues in the treatment of clinical depression, *IMA Journal of Mathematics Applied in Medicine & Biology* 12 (3–4), 369–385.

56. Schappert and Nelson, National ambulatory medical care survey.

57. J. H. Thakore and D. N. John, 1996, Prescription of antidepressants by general practitioners: recommendations by FHSAS and health boards, *British Journal of General Practice* 46 (407), 363–364.

58. Ibid.

59. R. E. Ferner and D. K. Scott, 1994, Whatalotwegot—the messages in drug advertisements, *British Medical Journal* 309 (6970), 1734–1736.

60. Pear, Spending on prescription drugs increases by almost 19 percent.

61. M. S. Wilkes et al., 1992, Pharmaceutical advertisements in leading medical journals: experts' assessments, *Annals of Internal Medicine* 116 (11), 912–919.

62. Lisa Belkin, 2001, Prime time pushers, *Mother Jones* (March–April).

6. Welfare, Cooperation, and Evolution

1. F. D. Parker and J. B. Welch, 1991, Field comparisons of attractants for the screw worm fly (Diptera, *Calliphoridae*) in a tropical dry forest in Costa Rica, *Journal of Economic Entomology* 84 (4), 1189–1195.

2. FAO, 1995, *Insecticidal Control of the New World Screwworm*. SCNA/INT/001/MUL (United Nations, New York); W. S. Patton, 1922, Notes on the myiasis producing Diptera of man and animals, *Bulletin of Entomological Research* 12, 239–261; D. B. Thomas and R. L. Mangan, 1989, Oviposition and wound-visiting behaviour of the screwworm fly, *Cochliomyia hominivorax* (Diptera: Calliphoridae), *Annals of the Entomological Society of America* 82, 526–534.

3. John Locke, 1693, *Some Thoughts Concerning Education* (A. and J. Churchill, London).

4. Adam Smith, 1933, *The Wealth of Nations* (J. M. Dent & Sons, London).

5. Robert H. Frank, 1988, *Passions within Reason: The Strategic Role of the Emotions* (Norton, New York).

6. Frank, *Passions within Reason*.

7. Love is not the same as lust,
   lust needs play love needs trust,
   one night stands and summer flings
   commitment, marriage, engagement rings
   love is the victor and will stand the time
   lust in a flash almost a crime. (Gabby Smith)

8. Derek Freeman has written extensively on the ruse that was perpetrated on Mead by two of her primary informants, Fa´apua'a Fa´amū and Galea´I Poumele. See Derek Freeman, 1983, *Margaret Mead and Samoa: The Making and Unmaking of an Anthropological Myth* (Harvard University Press, Cambridge, Mass.); Derek Freeman, 1999, *The Fateful Hoaxing of Margaret Mead: A Historical Analysis of Her Samoan Research* (Westview Press, Boulder).

9. Margaret Mead, 1928, *Coming of Age in Samoa; A Psychological Study of Primitive Youth for Western Civilisation* (W. Morrow, New York).

10. Freeman, *Margaret Mead and Samoa;* Freeman, *The Fateful Hoaxing of Margaret Mead*.

11. Charles Darwin, 1871, *The Descent of Man, and Selection in Relation to Sex* (John Murray, London).

12. Ernst Mayr, 1988, *Toward a New Philosophy of Biology: Observations of an Evolutionist* (Harvard University Press, Cambridge, Mass.).

13. E. O. Wilson (1975, *Sociobiology: The New Synthesis* [Harvard University Press, Cambridge, Mass.]) used the example of the profound social change that occurred in Ireland in response to the decimation of the potato crop in 1845 and 1846.

Between 1840 and 1911, the population of Ireland decreased from 8,200,000 to a staggering 4,400,000 due to disease, starvation, and emigration (R. F. Foster, 1988, *Modern Ireland, 1600–1972* [A. Lane, London]). As many as a quarter-million people left Ireland each year, and the famine-related death counts are confirmed at around 750,000 and estimated at about 1,500,000. Another stunning example of rapid cultural change occurred in Japan at the end of World War II. The occupation of Japan by the Allied powers started in August 1945 and ended in April 1952. Japan basically lost all the territory seized after 1894. In addition, the Kuril Islands were occupied by the Soviet Union, and the Ryukyu Islands, including Okinawa, were controlled by the United States. A new constitution went into effect in 1947: the emperor lost all political and military power and was made solely the symbol of the state. Universal suffrage was introduced, and human rights were guaranteed. Japan was also forbidden ever to lead a war again or to maintain a standing army. Furthermore, Shinto and the state were clearly separated.

14. Commission of the naturalistic fallacy would lead one to conclude that, since there have always been wars, the bombing in Serbia was perfectly natural, or since there is an unequal distribution of wealth in our society and those with the greatest fitness are likely to have the greatest resources, there is little reason to feed the poor.

15. G. E. Moore, 1922, *Principia Ethica* (Cambridge University Press, Cambridge).

16. The air crib was a large, soundproof, germ-free, air-conditioned box designed as a mechanical baby tender that provided an optimal environment for child growth during the first two years of life. The Skinner box was an experimental device used to train laboratory animals to perform complex and sometimes quite exceptional actions. The Skinner box has been widely used in pharmacological research to quantify the effects of drugs on behavior.

17. Behaviorism is the idea that behavior is governed by one's internal physiological state, which responds to changes in the environment.

18. B. F. Skinner, 1948, *Walden Two* (Macmillan, New York). Operant conditioning is a kind of learning that occurs in response to stimuli.

19. Charles R. Darwin, 1859, *On the Origin of Species by Means of Natural Selection* (J. Murray, London).

20. Ibid.

21. President George W. Bush at a South Carolina oyster roast, as quoted in the *Financial Times*, 14 January 2000.

22. Robert L. Trivers, 1971, The evolution of reciprocal altruism, *The Quarterly Review of Biology* 46, 35–57.

23. W. D. Hamilton, 1964, The genetical evolution of social behaviour, parts 1 and 2. *Journal of Theoretical Biology* 7, 1–52.

24. I. Seibold and A. J. Helbig, 1995, Evolutionary history of New and Old World vultures inferred from nucleotide sequences of the mitochondrial cytochrome B gene, *Philosophical Transactions of the Royal Society of London, Series B—Biological Sciences* 350 (1332), 163–178; M. Wink, 1995, Phylogeny of Old and New World vultures (Aves: Accipitridae and Cathartidae) inferred from nucleotide sequences of the mitochondrial cytochrome B gene, *Zeitschrift für Naturforschung C-Journal of Biosciences* 50 (11–12), 868–882. The general idea is that animals from

very different lineages can converge on similar adaptations to particular environmental conditions. Classic examples of convergence come from two groups of mammals, the placentals and the marsupials. Look at the evolution of placental and marsupial carnivores. The saber-toothed marsupial carnivore *Thylacosmilus*, which lived in South America in the Pliocene, is remarkably similar in morphology to the saber-toothed placental carnivore *Smilodon* from the Pleistocene in North America. M. Ridley, 1996, *Evolution* (Blackwell Science, Cambridge).

25. These data are used by some extremists to call for the abandonment of research on any primate because of our genetic closeness.

26. J.B.S. Haldane, 1955, Population genetics, *New Biology* 18, 34–51.

27. More recently, the traditional acquisition of kinship information has fallen out of favor with the rise of postmodern anthropology. The recognition of kinship is an intrinsic part of the human experience, but according to postmodern dogma, it makes little sense to try to map out any sort of collective understanding of biological and social relations since they are subject to individual interpretation.

28. Debra S. Judge, 1995, American legacies and the variable life histories of women and men, *Human Nature* 6 (4), 291–323; E. O. Smith, 1987, Deception and evolutionary biology, *Cultural Anthropology* 2, 50–64.

29. Christophe Boesch, 1996, Social grouping in Tai chimpanzees, in William C. McGrew and Linda F. Marchant, eds., *Great Ape Societies* (Cambridge University Press, Cambridge), pp. 101–113; Kenji Kawanaka, 1982, Further studies on predation by chimpanzees of the Mahale Mountains, *Primates* 23 (3), 364–384; C. B. Stanford et al., 1994, Patterns of predation by chimpanzees on red colobus monkeys in Gombe National Park, 1982–1991, *American Journal of Physical Anthropology* 94, 213–228; G. Teleki, 1973, *The Predatory Behavior of Wild Chimpanzees* (Bucknell University Press, Lewisburg, Pa.); S. Uehara, 1986, Sex and group differences in feeding on animals by wild chimpanzees in the Mahale Mountains National Park, Tanzania, *Primates* 27, 1–13; Shigeo Uehara, 1997, Predation on mammals by the chimpanzee (*Pan troglodytes*), *Primates* 38 (2), 193–214; R. W. Wrangham and E. Van-Zinnicq-Bergmann-Riss, 1990, Rates of predation on mammals by Gombe chimpanzees, 1972–1975, *Primates* 31, 157–170.

30. L. A. Dugatkin and M. Mesterton-Gibbons, 1996, Cooperation among unrelated individuals: reciprocal altruism, by-product mutualism and group selection in fishes, *Biosystems* 37 (1–2), 19–30; J. Grinnell et al., 1995, Cooperation in male lions: kinship, reciprocity or mutualism? *Animal Behaviour* 49 (1), 95–105; A. S. Grutter and R. Poulin, 1998, Cleaning of coral reef fishes by the wrasse *Labroides dimidiatus*: influence of client body size and phylogeny, *Copeia* 1, 120–127; M. I. Marcus, 1985, Feeding associations between capybaras and jacanas: a case of interspecific grooming and possibly mutualism, *Ibis* 127 (2), 240–243; D. McClearn, 1992, The rise and fall of a mutualism? Coatis, tapirs, and ticks on Barro Colorado Island, Panama, *Biotropica* 24 (2a), 220–222; Paul N. Newton, 1989, Associations between langur monkeys (*Presbytis entellus*) and chital deer (*Axis axis*): chance encounters or a mutualism? *Ethology* (formerly *Zeitschrift für Tierpsychologie*) 83 (2), 89–120; P. Weeks, 1999, Interactions between red-billed oxpeckers, *Buphagus erythrorhynchus*, and domestic cattle, *Bos taurus*, in Zimbabwe, *Animal Behaviour* 58

(6), 1253–1259; P. Weeks, 2000, Red-billed oxpeckers: vampires or tickbirds? *Behavioral Ecology* 11 (2), 154–160.

31. Trivers, The evolution of reciprocal altruism.

32. Ibid.

33. Ronald M. Nowak and Ernest P. Walker, 1991, *Walker's Mammals of the World*, vol. 1 (Johns Hopkins University Press, Baltimore).

34. Lisa K. DeNault and Donald A. McFarlane, 1995, Reciprocal altruism between male vampire bats, *Desmodus rotundus, Animal Behaviour* 49 (3), 855–856; Gerald S. Wilkinson, 1984, Reciprocal food sharing in the vampire bat, *Nature* 308 (5955), 181–184; Gerald S. Wilkinson, 1988, Reciprocal altruism in bats and other mammals, *Ethology & Sociobiology* 9 (2–4), 85–100.

35. Matthew E. Gompper et al., 1997, Genetic relatedness, coalitions and social behaviour of white-nosed coatis, *Nasua narica, Animal Behaviour* 53 (4), 781–797; C. Packer, 1977, Reciprocal altruism in *Papio anubis, Nature* 265, 441–443; Dugatkin and Mesterton-Gibbons, Cooperation among unrelated individuals.

36. Greg Moran, 1984, Vigilance behaviour and alarm calls in a captive group of meerkats, *Suricata suricatta, Zeitschrift für Tierpsychologie* 65 (3), 228–240.

37. T. H. Clutton-Brock et al., 1999, Selfish sentinels in cooperative mammals, *Science* 284 (5420), 1640–1644.

38. National Traffic Safety Board, 1982, *Air Florida, Inc., Boeing 737–222, N62AF, Collision with 14th Street Bridge, near Washington National Airport, Washington, D.C.,* 13 January 1982, AAR–82–08.

39. It is also true that Dr. Albert Sabin did not patent the oral polio vaccine and did not receive any compensation directly from its development. See Valerie Manda, 2001, Sabin and OPV, Personal communication, 7 March.

40. Whoever gives *tzedakah* to the poor with a sour expression and in a surly manner, even if he gives a thousand gold pieces, loses his merit. One should instead give cheerfully and joyfully, and empathize with him in his sorrow as it is said (Job 30:25), "Did I not cry for him whose day is difficult? Did my soul not grieve for the poor?" Speak to him with compassion and comfort as it is said (Job 29:13).

41. W. P. Dittus and S. M. Ratnayeke, 1989, Individual and social behavioral responses to injury in wild toque macaques (*Macaca sinica*), *International Journal of Primatology* 10, 215–234.

42. P. L. Whitten, 2001, Personal communication, Atlanta, Ga.

43. O. Anne Rasa, 1976, Invalid care in the dwarf mongoose (*Helogale undulata rufula*), *Zeitschrift für Tierpsychologie* 42 (4), 337–342; O. Anne Rasa, 1983, A case of invalid care in wild dwarf mongooses, *Zeitschrift für Tierpsychologie* 62 (3), 235–240.

44. D. O. Hebb, 1949, Temperament in chimpanzees I: method of analysis, *Journal of Comparative and Physiological Psychology* 42, 192–206; G. R. Stephenson, 1967, Cultural acquisition of a specific learned response among rhesus monkeys, in D. Starck et al., eds., *Progress in Primatology* (G. Fischer, Stuttgart), pp. 279–288.

45. A familiar example to many women is the injury and deformation inflicted on the bones in the foot as the result of wearing high heels. The well-known bunion and

the "pump bump" are striking testimonials to the ease with which bone can be re-modeled. E. O. Smith and W. S. Helms, 1999, Natural selection and high heels, *Foot and Ankle International* 20 (1), 55–57; E. O. Smith, 1999, High heels and evolution, *Psychology, Evolution, and Gender* 1 (3), 245–277.

46. R. S. Solecki, 1971, *Shanidar: The First Flower People* (Knopf, New York).

47. Arlette Leroi-Gourhan, 1975, The flowers found with Shanidar IV, a Nean-derthal burial in Iraq, *Science* 190 (4214), 562–564.

48. Erik Trinkaus and Pat Shipman, 1994, *The Neandertals: Of Skeletons, Scien-tists, and Scandal* (Vintage Books, New York).

49. D. W. Frayer et al., 1987, Dwarfism in an adolescent from the Italian late Up-per Palaeolithic, *Nature* 330 (6143), 60–62; D. W. Frayer et al., 1988, A case of chon-drodystrophic dwarfism in the Italian late Upper Paleolithic, *American Journal of Physical Anthropology* 75 (4), 549–565.

50. Solecki, *Shanidar*; Erik Trinkaus, 1983, *The Shanidar Neandertals* (Academic Press, New York); Milford H. Wolpoff, 1999, *Paleoanthropology* (McGraw-Hill, New York).

51. Katherine Dettwyler (1991, Can paleopathology provide evidence for "com-passion"? *American Journal of Physical Anthropology* 84 [4], 375–384) has argued that the fact that individuals have survived traumatic injury does not definitely demonstrate altruistic aid and support. She cites the classic case of Faben, an adult male chimpanzee at the Gombe Stream Reserve who contracted polio and lost the use of his right arm but was to survive for a number of years. See Jane Goodall, 1971, *In the Shadow of Man* (Houghton Mifflin, Boston); Jane Goodall, 1986, Social re-jection, exclusion, and shunning among the Gombe chimpanzees, *Ethology & So-ciobiology* 7 (3–4), 227–236.

52. For much of this discussion I have relied on V. Bernhard et al., 1992, *First-hand America: A History of the United States* (Brandywine Press, St. James, N.Y.).

53. Almshouses have existed in the United Kingdom for over one thousand years. The first almshouse was founded by King Athelstan in York in A.D. 937. Early almshouses were called "hospitals" because their goal was to offer hospitality and shelter for the poor and needy. Almshouses were frequently associated with the Church, but, after the dissolution of the monasteries by King Henry VIII in the six-teenth century, many almshouses became endowed and managed by city livery companies, town mayors, or wealthy landowners for the benefit of local people or those in a particular trade. In the nineteenth century, many industrialists founded almshouse charities for their retired workers. Almshouse Association, 2001, Almshouses organization, *http://www.almshouses.org/*.

54. An excellent biography of Long can be found in T. Harry Williams, 1981, *Huey Long* (Vintage, New York).

55. Itta Bena, Mississippi, is located about ninety miles north of the state capital of Jackson. It is situated about five miles northwest of Quito and about ten miles west of Greenwood in the heart of the Mississippi Delta.

56. A "zero-sum" game is simply a win-lose game such as tic-tac-toe. For every winner, there's a loser. If I win, you lose. Nonzero-sum games allow for cooperation.

57. Robert M. Axelrod, 1984, *The Evolution of Cooperation* (Basic Books, New York).

58. Axelrod noted that mutual defection is preferred to unilateral restraint and unilateral restraint by the opposition is better than mutual cooperation. In addition, the reward for mutual restraint is preferred to the outcome of mutual punishment, since both participants would suffer for little or no gain. These situations set up the essential conditions for Prisoner's Dilemma. Ibid.

59. It is interesting that many of the ideas discussed in this chapter are not news at all to much of the business community. Altruism has become a highly developed bit of the psychological warfare that advertisements wage on the American public every day. Americans are admonished to "give 'til it hurts," to support famine/earthquake/disaster relief in the population of your choice, and the like. Americans, like people in other developed countries, are urged to give resources to those less fortunate than themselves. The payoff can come in several different currencies, but, no matter what the currency, the most successful of these advertising campaigns are the ones that specify most clearly the payoff to the altruist. For many, the payoff for altruistic behavior is a tax deduction or a chance to win a prize: a new car, a box seat at a sporting event, and so forth. On a somewhat less crass note, the payoff may be in enhanced self-esteem and self-worth. We may also feel that, by being altruistic, we are currying future favors in some fashion. Religious people may engage in altruistic behaviors because they feel that they are improving their chances of a good afterlife. Those not so religious may think that, by being good, they will be rewarded in their own time of need. One part of our day-to-day experience that seems to bear this future payoff idea is in driving practices. Be nice to the person trying to merge into traffic, and someone will be nice to you. Always treat others as you want to be treated, and you will be rewarded with good parking places.

60. Trustees of Social Security and Medicare Boards, 2000, Status of the Social Security and Medicare programs: a summary of the 2000 annual reports, *http://www.ssa.gov/OACT/TRSUM/trsummary.html*.

61. When a government transfers money from one societal group to another, it is called a transfer payment or "entitlement." For example, the government can tax the general public in order to give money to the poor in the form of welfare, or the government can transfer money from workers paying Social Security taxes to the elderly who are entitled to receive Social Security benefits. Other forms of transfer payments in the United States include the commerce and housing credit, which gives money to people who cannot provide housing for themselves, and Medicare, which provides health care for the elderly.

References

Ablaza, V. J., M. R. Jones, M. K. Gingrass, J. Fisher, and G. P. Maxwell. 1998. Ultrasound assisted lipoplasty—part 1: an overview for nurses. *Plastic Surgical Nursing* 18 (1), 13–15.

Alcock, J. 1998. *Animal Behavior*. Sinauer, Sunderland.

Alderman, M. H., H. Cohen, and S. Madhavan. 1998. Dietary sodium intake and mortality: the national health and nutrition examination survey (NHANES I). *Lancet* 351 (9105), 781–785.

Alexander, R. D., J. L. Hoogland, R. D. Howard, K. M. Noonan, and P. W. Sherman. 1979. Sexual dimorphisms and breeding systems in pinnipeds, ungulates, primates, and humans. In N. A. Chagnon and W. Irons, eds., *Evolutionary Biology and Human Social Behavior: An Anthropological Perspective*. Duxbury Press, North Scituate, Mass., pp. 402–435.

Almshouse Association. 2001. Almshouses organization. *http://www.almshouses.org/*.

American Academy of Cosmetic Surgery. 2001. Statistics of cosmetic procedures. *http://www.cosmeticsurgery.org/consumer/stats/2000statistics*.

American Psychiatric Association. 1994. *Diagnostic and Statistical Manual of Mental Disorders, DSMIV*. American Psychiatric Association, Washington, D.C.

American Society of Plastic Surgeons. 2001. 2000 average surgeon/physician fees, cosmetic surgery. *http://www.plasticsurgery.org/mediactr/average.pdf*.

Andreasen, N. C., and G. Winokur. 1979. Secondary depression: familial, clinical, and research perspectives. *American Journal of Psychiatry* 136 (1), 62–66.

Ardrey, Robert. 1966. *The Territorial Imperative: A Personal Inquiry into the Animal Origins of Property and Nations*. Atheneum, New York.

Armelagos, George J. 1987. Biocultural aspects of food choice. In M. Harris and Eric B. Ross, eds., *Food and Evolution: Toward a Theory of Human Food Habits*. Temple University Press, Philadelphia, pp. 579–594.

Aronson, J. K., and M. Hardman. 1992. ABC of monitoring drug therapy: patient compliance. *British Medical Journal* 305 (6860), 1009–1011.

———. 1992. ABC of monitoring drug therapy: why monitor drug therapy? *British Medical Journal* 305 (6859), 947–948.

Arthey, S., and V. A. Clarke. 1995. Suntanning and sun protection: a review of the psychological literature. *Social Science & Medicine* 40 (2), 265–274.

Associated Press. 2001. Indian beef protesters raid McDonald's. Associated Press, 6 May.

Austin, Elizabeth. 1999. Marks of mystery. *Psychology Today* 32, 46–50.

Axelrod, Robert M. 1984. *The Evolution of Cooperation*. Basic Books, New York.

Bailey, Stephen M. 1982. Absolute and relative sex differences in body composition.

In Roberta L. Hall, ed., *Sexual Dimorphism in Homo sapiens: A Question of Size.* Praeger, New York, pp. 363–390.

Bannon, L. 1997. She reinvented Barbie; now can Jill Bardo do the same for Mattel? *Wall Street Journal*, 5 March, final edition, section A, p. 1.

Barbieri, Gian Paolo. 1998. *Tahiti Tattoos.* Taschen, New York.

Beauchamp, G. K., M. Bertino, D. Burke, and K. Engelman. 1990. Experimental sodium depletion and salt taste in normal human volunteers. *American Journal of Clinical Nutrition* 51 (5), 881–889.

Belkic, K., R. Emdad, and T. Theorell. 1998. Occupational profile and cardiac risk: possible mechanisms and implications for professional drivers. *International Journal of Occupational Medicine & Environmental Health* 11 (1), 37–57.

Belkin, Lisa. 2001. Prime time pushers. *Mother Jones* (March–April).

Bellafante, Ginia. 2001. The way we live: 8–12–01; tan is the new black. *New York Times Magazine*, 12 August, pp. 10–12.

Bennett, William, and Joel Gurin. 1982. *The Dieter's Dilemma.* Basic Books, New York.

Bereczkei, Tamas, Silvia Voros, Agnes Gal, and Laszlo Bernath. 1997. Resources, attractiveness, family commitment: reproductive decisions in human mate choice. *Ethology* 103 (8), 681–699.

Berman, M. E., J. I. Tracy, and E. F. Coccaro. 1997. The serotonin hypothesis of aggression revisited. *Clinical Psychology Review* 17 (6), 651–665.

Bermant, G. 1976. Sexual behavior: hard times with the Coolidge effect. In M. H. Siegel and H. P. Zeigler, eds., *Psychological Research: The Inside Story.* Harper and Row, New York, pp. 76–103.

Bernes, S. M., and A. M. Kaplan. 1994. Evolution of neonatal seizures. *Pediatric Clinics of North America* 41 (5), 1069.

Bernhard, V., D. Burner, E. Fox-Genovese, E. D. Genovese, J. McClymer, and F. McDonald. 1992. *Firsthand America: A History of the United States.* Brandywine Press, St. James, N.Y.

Berscheid, E., and E. Walster. 1983. Physical attractivenes. In L. Berkowitz, ed., *Advances in Experimental Social Psychology.* Academic Press, New York, pp. 157–215.

Bertino, M., G. K. Beauchamp, D. R. Riskey, and K. Engelman. 1981. Taste perception in three individuals on a low sodium diet. *Appetite* 2 (1), 67–73.

Blackburn, Mark. 1999. *Tattoos from Paradise: Traditional Polynesian Patterns.* Schiffer, Atglen, Pa.

Blair, S. N., H. W. Kohl, III, C. E. Barlow, R. S. Paffenbarger, Jr., L. W. Gibbons, and C. A. Macera. 1995. Changes in physical fitness and all-cause mortality: a prospective study of healthy and unhealthy men. *Journal of the American Medical Association* 273 (14), 1093–1098.

Bloch, David. 1999. Salt and gold. *Journal of Salt History* 7, 9.

Bloch, P. H., and M. L. Richins. 1992. You look "mahvelous": the pursuit of beauty and the marketing concept. *Psychology & Marketing* 9 (1), 3–15.

Blumberg, P., and L. P. Mellis. 1980. Medical students' attitudes toward the obese and the morbidly obese. *International Journal of Eating Disorders* 4, 169–175.

Boesch, Christophe. 1996. Social grouping in Tai chimpanzees. In William C. McGrew and Linda F. Marchant, eds., *Great Ape Societies*. Cambridge University Press, Cambridge, pp. 101–113.

Bogart, Leo. 1972. Warning: The surgeon general has determined that TV violence is moderately dangerous to your child's mental health. *Public Opinion Quarterly* 36 (4), 491–521.

Bohannan, Laura, and Paul Bohannan. 1953. *The Tiv of Central Nigeria*. International African Institute, London.

Bramble, D. J. 1995. Antidepressant prescription by British child psychiatrists: practice and safety issues. *Journal of the American Academy of Child & Adolescent Psychiatry* 34 (3), 327–331.

Brandes, L. J. 1992. Depression, antidepressant medication, and cancer. *American Journal of Epidemiology* 136 (11), 1414–1417.

Brandes, L. J., and M. Cheang. 1995. Letter regarding "response to antidepressants and cancer: cause for concern?" *Journal of Clinical Psychopharmacology* 15 (1), 84–85.

Brink, P. J. 1989. The fattening room among the Annang of Nigeria. *Medical Anthropology* 12 (1), 131–143.

British Broadcasting Corporation. 2001. Obesity gene pinpointed. *http://news.bbc.co.uk/hi/english/health/newsid_1484000/484659.stm*.

Broadstock, M., R. Borland, and G. Gason. 1992. Effects of suntan on judgements of healthiness and attractiveness by adolescents. *Journal of Applied Social Psychology* 22 (2), 157–172.

Brown, Peter J., and Melvin J. Konner. 1987. An anthropological perspective on obesity. *Annals of the New York Academy of Sciences* 499, 29–46.

Brown, Rebecca. 1993. Barbie comes out. In Lucinda Ebersole and Richard Peabody, eds., *Mondo Barbie*. St. Martin's, New York, pp. 152–165.

Brownell, K. D., and M. A. Napolitano. 1995. Distorting reality for children: body size proportions of Barbie and Ken dolls. *International Journal of Eating Disorders* 18 (3), 295–298.

Brumberg, J. J. 1988. *Fasting Girls: The Emergence of Anorexia Nervosa as a Modern Disease*. Harvard University Press, Cambridge, Mass.

Bureau of Labor Statistics. 2000. Occupational outlook handbook, 2000–01: service occupations: barbers, cosmetologists and related workers. *http://stats.bls.gov/oco/ocos169.htm*.

Buss, David M. 1991. Do women have evolved mate preferences for men with resources? A reply to Smuts. *Ethology & Sociobiology* 12 (5), 401–408.

———. 1999. *Evolutionary Psychology: The New Science of the Mind*. Allyn & Bacon, Boston.

Buydens-Branchey, L., M. Branchey, J. Hudson, and P. Fergeson. 2000. Low HDL cholesterol, aggression and altered central serotonergic activity. *Psychiatry Research* 93 (2), 93–102.

Cades, Hazel Rawson. 1927. *Any Girl Can Be Good-Looking*. D. Appleton, New York.

Caine, Mead T. 1977. The economic activities of children in a village in Bangladesh. *Population and Development Review* 13 (3), 201–227.

Campfield, L. A., F. J. Smith, M. Rosenbaum, and J. Hirsch. 1996. Human eating: evidence for a physiological basis using a modified paradigm. *Neuroscience & Biobehavioral Reviews* 20 (1), 133–137.

Cash, T. F., and K. L. Hicks. 1990. Being fat versus thinking fat: relationships with body image, eating behaviors and well-being. *Cognitive Therapy and Research* 14, 327–341.

Cash, T. F., B. A. Winstead, and L. H. Janda. 1986. Body image survey report: the great American shape up. *Psychology Today* 20, 30–37.

Centers for Disease Control and Prevention. 2001. Facts and statistics about skin cancer. *http://www.cdc.gov/chooseyourcover/skin.htm.*

Chernin, Kim. 1985. *The Hungry Self: Women, Eating and Identity.* Times Books, New York.

Children's Market Services, Inc. 2001. Strategic planners in the youth market. *http://www.kidtrends.com.*

Chisholm, Michael, and David Marshall Smith. 1990. *Shared Space, Divided Space: Essays on Conflict and Territorial Organization.* Unwin Hyman, London.

Chutkow, Paul. 1999. The real Vargas: one hundred years after his birth, Alberto Vargas is still regarded as "the King of Pin-up Art," and therein hangs an exquisite love story—and a terrible tragedy. *Cigar Aficionado* 9 (1), 23–24.

Clary, Mike. 2001. Simpson faces road rage charges; law: football hall of famer is arrested on burglary and battery counts in Miami incident. He says he welcomes felony case. *Los Angeles Times*, February 10, home edition, section A; part 1, p. 3.

Clutton-Brock, T. H., and G. R. Iason. 1986. Sex ratio variation in mammals. *Quarterly Review of Biology* 61, 339–374.

Clutton-Brock, T. H., M. J. O'Riain, P.N.M. Brotherton, D. Gaynor, R. Kansky, A. S. Griffin, and M. Manser. 1999. Selfish sentinels in cooperative mammals. *Science* 284 (5420), 1640–1644.

Cohen, Mark Nathan. 1989. *Health and the Rise of Civilization.* Yale University Press, New Haven.

Cohen, Mark Nathan, and George J. Armelagos, eds. 1984. *Paleopathology at the Origins of Agriculture.* Academic Press, New York.

Comstock, George. 1986. Television and film violence. In Steven J. Apter and Arnold P. Goldstein, eds., *Youth Violence: Programs and Prospects.* Pergamon Press, Oxford, pp. 178–218.

Connell, Dominic, and Matthew Joint. 1996. *Driver Aggression.* MCIT Road Safety Unit, London.

Conyngton, T. 1899. Motor carriages and street paving. *Scientific American* 48 (supplement no. 1226), 196.

Cooper, Richard S., and Charles N. Rotimi. 1994. Hypertension in populations of West African origin: is there a genetic predisposition? *Journal of Hypertension* 12 (3), 215–227.

Cooper, Richard S., Charles N. Rotimi, Susan L. Ataman, Daniel L. McGee, Babatunde Osotimehin, Solomon Kadiri, Walinjom Muna, Samuel Kingue, Henry Fraser, Terrence Forrester, Franklyn Bennett, and Rainford Wilks. 1997.

Hypertension prevalence in seven populations of African origin. *American Journal of Public Health* 87 (2), 160–168.

Cooper, Richard S., Charles N. Rotimi, and Ryk Ward. 1999. The puzzle of hypertension in African-Americans. *Scientific American* 280 (2), 56–63.

Cravens, Hamilton. 1990. Establishing the science of nutrition at the USDA: Ellen Swallow Richards and her allies. *Agricultural History* 64 (2), 122–133.

Cross-National Collaborative Group. 1992. The changing rate of major depression: cross-national comparisons. *Journal of the American Medical Association* 268 (21), 3098–3105.

Cunningham, Michael R., Alan R. Roberts, Anita P. Barbee, Perri B. Druen, et al. 1995. "Their ideas of beauty are, on the whole, the same as ours": consistency and variability in the cross-cultural perception of female physical attractiveness. *Journal of Personality & Social Psychology* 68 (2), 261–279.

Darwin, Charles R. 1859. *On the Origin of Species by Means of Natural Selection.* John Murray, London.

———. 1871. *The Descent of Man, and Selection in Relation to Sex.* John Murray, London.

Davis, Caroline, and Etta Saltos. 1999. Dietary recommendations and how they have changed over time. In Elizabeth Frazäo, ed., *America's Eating Habits: Changes and Consequences.* Agriculture Information Bulletin no. 750. Food and Rural Economics Division, Economic Research Services, USDA, Washington, D.C., pp. 32–50.

DeJong, W., and R. E. Kleck. 1986. The social psychological effects of overweight. In P. Herman, M. P. Zanna, and E. T. Higgins, eds., *Physical Appearance, Stigma, and Social Behavior: The Ontario Symposium.* Vol. 3. Lawrence Erlbaum, Hillsdale, N.J., pp. 65–87.

DeNault, Lisa K., and Donald A. McFarlane. 1995. Reciprocal altruism between male vampire bats, *Desmodus rotundus. Animal Behaviour* 49 (3), 855–856.

Denton, Derek A. 1982. *The Hunger for Salt: An Anthropological, Physiological, and Medical Analysis.* Springer-Verlag, Berlin.

De Silva, Harry Reginald. 1942. *Why We Have Automobile Accidents.* J. Wiley & Sons, New York.

Desjarlais, Robert, Leon Eisenberg, Byron Good, and Arthur Kleinman. 1995. *World Mental Health: Problems and Priorities in Low-Income Countries.* Oxford University Press, Oxford.

Dettwyler, K. A. 1991. Can paleopathology provide evidence for "compassion"? *American Journal of Physical Anthropology* 84 (4), 375–384.

Dhurandhar, N. V., B. A. Israel, J. M. Kolesar, G. F. Mayhew, M. E. Cook, and R. L. Atkinson. 2000. Increased adiposity in animals due to a human virus. *International Journal of Obesity & Related Metabolic Disorders* 24 (8), 989–996.

Diamond, Jared M. 1987. The worst mistake in the history of the human race. *Discover* 8 (5), 64–66.

Dittus, W. P., and S. M. Ratnayeke. 1989. Individual and social behavioral responses to injury in wild toque macaques (*Macaca sinica*). *International Journal of Primatology* 10, 215–234.

Djawdan, M., and T. Garland. 1988. Maximal running speeds of bipedal and quadrupedal rodents. *Journal of Mammalogy* 69 (4), 765–772.

Dubovsky, S. L. 1994. Beyond the serotonin reuptake inhibitors: rationales for the development of new serotonergic agents. *Journal of Clinical Psychiatry* 55, 34–44.

Dugatkin, L. A., and M. Mesterton-Gibbons. 1996. Cooperation among unrelated individuals: reciprocal altruism, by-product mutualism and group selection in fishes. *Biosystems* 37 (1–2), 19–30.

Dyson Hudson, R., and E. A. Smith. 1978. Human territoriality: an ecological re-assessment. *American Anthropologist* 80, 21–41.

Eaton, S. B., and M. Konner. 1985. Paleolithic nutrition: a consideration of its nature and current implications. *New England Journal of Medicine* 312, 283–289.

———. 1997. Paleolithic nutrition revisited: a twelve-year retrospective on its nature and implications. *European Journal of Clinical Nutrition* 51 (4), 207–216.

Eaton, S. B., M. Shostak, and M. Konner. 1988. *The Paleolithic Prescription: A Program of Diet and Exercise and a Design for Living.* Harper & Row, New York.

Ebensten, Hanns. 1953. *Pierced Hearts and True Love.* D. Verschoyle, London.

Ebersole, Lucinda, and Richard Peabody, eds. 1993. *Mondo Barbie.* St. Martin's, New York.

Eccleston, D. 1993. The economic evaluation of antidepressant drug-therapy. *British Journal of Psychiatry* 163, 5–6.

Edney, Julian J. 1974. Human territoriality. *Psychological Bulletin* 81 (12), 959–975.

Edwards, D. H., and E. A. Kravitz. 1997. Serotonin, social status and aggression. *Current Opinion in Neurobiology* 7 (6), 812–819.

Ellis, Albert. 1991. Suggestibility, irrational beliefs, and emotional disturbance. In John F. Schumaker, ed., *Human Suggestibility: Advances in Theory, Research, and Application.* Taylor & Francis/Routledge, Florence, Ky., pp. 309–325.

Ellison, P. A., J. M. Govern, H. L. Petri, and M. H. Figler. 1995. Anonymity and aggressive driving behavior: a field study. *Journal of Social Behavior and Personality* 10 (1), 265–272.

Ellison-Potter, Patricia, Paul Bell, and Jerry Deffenbacher. 2001. The effects of trait driving anger, anonymity, and aggressive stimuli on aggressive driving behavior. *Journal of Applied Social Psychology* 31 (2), 431–443.

Emerson, Sam. 1984. The last word on Michael. *People* (November/December), 97.

Encyclopædia Britannica. 2001. Bile. *http://www.britannica.com/eb/article?eu=81340&tocid=0&query=bile.*

———. 2001. Body modification and mutilation. *http://www.britannica.com/eb/article?eu=82520&tocid=8270&query=cranial%20defomation.*

———. 2001. Duel. *http://www.britannica.com/eb/article?eu=31895&tocid=0&query=duelling%20scars.*

Encyclopedia Infoplease. 2001. Gutta percha. *http://www.infoplease.com/ce6/sci/A0822209.html.*

Fallon, A. 1990. Culture in the mirror: sociocultural determinants of body image. In T. F. Cash and T. Pruzinsky, eds., *Body Images: Development, Deviance, and Change.* Guilford Press, New York, pp. 80–109.

FAO. 1995. *Insecticidal Control of the New World Screwworm*. SCNA/INT/ 001/MUL. United Nations, New York.

Farnham. Alan. 1996. Male vanity: you're so vain I bet you think this story's about you: American men, refusing to age gracefully, are spending $9.5 billion a year on face-lifts, hairpieces, and makeup. *Fortune* 134 (5), 66–73.

Feldman, W., E. Feldman, and J. T. Goodman. 1988. Culture versus biology: children's attitudes toward thinness and fatness. *Pediatrics* 81 (2), 190–194.

Felson, Richard B. 1996. Mass media effects on violent behavior. *Annual Review of Sociology* 22, 103–128.

Ferner, R. E., and D. K. Scott. 1994. Whatalotwegot—the messages in drug advertisements. *British Medical Journal* 309 (6970), 1734–1736.

Ferris, C. F., T. Stolberg, and Y. Delville. 1999. Serotonin regulation of aggressive behavior in male golden hamsters (*Mesocricetus auratus*). *Behavioral Neuroscience* 113 (4), 804–815.

Fisher, R. L., and S. Fisher. 1996. Antidepressants for children: is scientific support necessary? *Journal of Nervous & Mental Disease* 184 (2), 99–102.

Flaherty, Michael G. 1999. *A Watched Pot: How We Experience Time*. New York University Press, New York.

Flink, James J. 1975. *The Car Culture*. MIT Press, Cambridge, Mass.

Fodor, J. George, Beverly Whitmore, Frans H. Leenen, and Pierre Larochelle. 1999. Recommendations on dietary salt. *Canadian Medical Association Journal* 160 (9 supplement), S29–S34.

Fosarelli, P. D. 1984. Televison and children: a review. *Journal of Developmental and Behavioral Pediatrics* 5, 30–37.

Foster, R. F. 1988. *Modern Ireland, 1600–1972*. Viking Penguin, London.

Frank, Robert H. 1988. *Passions within Reason: The Strategic Role of the Emotions*. Norton, New York.

Frayer, D. W., W. A. Horton, R. Macchiarelli, and M. Mussi. 1987. Dwarfism in an adolescent from the Italian late Upper Palaeolithic. *Nature* 330 (6143), 60–62.

Frayer, D. W., R. Macchiarelli, and M. Mussi. 1988. A case of chondrodystrophic dwarfism in the Italian late Upper Paleolithic. *American Journal of Physical Anthropology* 75 (4), 549–565.

Freedman, R. J. 1984. Reflections on beauty as it relates to health in adolescent females. *Women & Health* 9, 29–45.

Freeman, Derek. 1983. *Margaret Mead and Samoa: The Making and Unmaking of an Anthropological Myth*. Harvard University Press, Cambridge, Mass.

———. 1999. *The Fateful Hoaxing of Margaret Mead: A Historical Analysis of Her Samoan Research*. Westview Press, Boulder.

Friedman, R. B. 1986. Fad diets: evaluation of five common types. *Postgraduate Medicine* 79 (1), 249–255.

Fumento, Michael. 1998. "Road rage" versus reality. *The Atlantic Monthly* 282 (2), 12–17.

Gangestad, S. W., R. Thornhill, and R. A. Yeo. 1994. Facial attractiveness, developmental stability, and fluctuating asymmetry. *Ethology and Sociobiology* 15, 73–85.

Garner, David M., Paul E. Garfinkel, Donald Schwartz, and Michael Thompson.

1980. Cultural expectations of thinness in women. *Psychological Reports* 47 (2), 483–491.

Garreau, Joel. 1991. *Edge City: Life on the New Frontier.* Doubleday, New York.

Gatesy, S. M., and A. A. Biewener. 1991. Bipedal locomotion: effects of speed, size and limb posture in birds and humans. *Journal of Zoology* 224, 127–148.

Gilbert, Steve. 2000. *The Tattoo History Source Book.* Juno Books, New York.

Gladwell, Malcolm. 2001. The trouble with fries; fast food is killing us. Can it be fixed? *The New Yorker,* 5 March, p. 52.

Gliddon, D. 1983. Through the looking glasses of physicians, dentists, and patients. *Perspectives in Biology and Medicine* 26, 451–458.

Goldberg, R. J. 1997. Antidepressant use in the elderly: current status of nefazodone, venlafaxine and moclobemide. *Drugs & Aging* 11 (2), 119–131.

Goldstein, D. J. 1995. Effects of third trimester fluoxetine exposure on the newborn. *Journal of Clinical Psychopharmacology* 15 (6), 417–420.

Goldstein, D. J., L. A. Corbin, and K. L. Sundell. 1997. Effects of first-trimester fluoxetine exposure on the newborn. *Obstetrics & Gynecology* 89 (5 part 1), 713–718.

Goldstein, D. J., K. L. Sundell, and L. A. Corbin. 1997. Birth outcomes in pregnant women taking fluoxetine. *New England Journal of Medicine* 336 (12), 872–873; discussion, 873.

Gompper, Matthew E., John L. Gittleman, and Robert K. Wayne. 1997. Genetic relatedness, coalitions and social behaviour of white-nosed coatis, *Nasua narica. Animal Behaviour* 53 (4), 781–797.

Goodall, Jane. 1971. *In the Shadow of Man.* Houghton Mifflin, Boston.

———. 1986. Social rejection, exclusion, and shunning among the Gombe chimpanzees. *Ethology & Sociobiology* 7 (3–4), 227–236.

Gorn, Gerald J., and Renee Florsheim. 1985. The effects of commercials for adult products on children. *Journal of Consumer Research* 11 (4), 962–967.

Gorn, Gerald J., and Marvin E. Goldberg. 1987. Television and children's food habits: a big brother/sister approach. In Michael E. Manley-Casimir and Carmen Luke, eds., *Children and Television: A Challenge for Education.* Praeger, New York, pp. 34–48.

Gortmaker, S. L., A. Must, J. M. Perrin, A. M. Sobol, and W. H. Dietz. 1993. Social and economic consequences of overweight in adolescence and young adulthood. *New England Journal of Medicine* 329 (14), 1008–1012.

Graham, Pita. 1994. *Maori Moko or Tattoo.* Bush Press, Auckland, N.Z.

Grammer, K., and R. Thornhill. 1994. Human (*Homo sapiens*) facial attractiveness and sexual selection: the role of symmetry and averageness. *Journal of Comparative Psychology* (formerly *Journal of Comparative and Physiological Psychology*) 108 (3), 233–242.

Grayson, Richard. 1993. Twelve-step Barbie. In Lucinda Ebersole and Richard Peabody, eds., *Mondo Barbie.* St. Martin's, New York, pp. 48–54.

Grazier, Frederick M., and Rudolph H. De Jong. 2000. Fatal outcomes from liposuction: census survey of cosmetic surgeons. *Plastic and Reconstructive Surgery* 105 (1), 436–446.

Greenberg, B. S., and C. K. Atkin. 1983. The portrayal of driving on television, 1975–1980. *Journal of Communication* 33 (2), 44–55.

Greenberg, R. P., R. F. Bornstein, M. J. Zborowski, S. Fisher, and M. D. Greenberg. 1994. A meta-analysis of fluoxetine outcome in the treatment of depression. *Journal of Nervous & Mental Disease* 182 (10), 547–551.

Gregerson, E. 1983. *Sexual Practices: The Story of Human Sexuality*. Franklin Watts, New York.

———. 1994. *The World of Human Sexuality: Behaviors, Customs, and Beliefs*. Irvington Publishers, New York.

Grodstein, F., R. Levine, L. Troy, T. Spencer, G. A. Colditz, and M. J. Stampfer. 1996. Three-year follow-up of participants in a commercial weight loss program: can you keep it off? *Archives of Internal Medicine* 156 (12), 1302–1306.

Haiken, Elizabeth. 1997. *Venus Envy: A History of Cosmetic Surgery*. Johns Hopkins University Press, Baltimore.

Hakkert, A. S., V. Gitelman, and A. K. Cohen. 2001. The evaluation of effects on drivers' behavior and accidents of concentrated general enforcement on interurban roads in Israel. *Accident Analysis & Prevention* 33 (1), 43–63.

Halaas, J. L., K. S. Gajiwala, M. Maffei, S. L. Cohen, B. T. Chait, D. Rabinowitz, R. L. Lallone, S. K. Burley, and J. M. Friedman. 1995. Weight-reducing effects of the plasma protein encoded by the obese gene. *Science* 269 (5223), 543–546.

Haldane, J.B.S. 1955. Population genetics. *New Biology* 18, 34–51.

Hamilton, W. D. 1964. The genetical evolution of social behaviour, parts 1 and 2. *Journal of Theoretical Biology* 7, 1–52.

Handler, Ruth, and Jacqueline Shannon. 1994. *Dream Doll: The Ruth Handler Story*. Longmeadow Press, Stamford, Conn.

Handy, Willowdean Chatterson. 1922. *Tattooing in the Marquesas*. The Honolulu Museum Press, Honolulu.

———. 1965. *Forever the Land of Men: An Account of a Visit to the Marquesas Islands*. Dodd Mead, New York.

Hanneman, R. L. 1996. Intersalt: hypertension rise with age revisited. *British Medical Journal* 312 (7041), 1283–1284; discussion, 1284–1287.

Hansard, T. C. 1806. *The Parliamentary History of England from the Earliest Period to the Year 1803, from Which Last-Mentioned Epoch It Is Continued Downwards in the Work Entitled "Hansard's Parliamentary Debates."* T. C. Hansard, London.

Harding, T. 2000. U.K.'s drivers overtake Europe for road rage. *The Daily Telegraph*, 15 May, p. 5.

Harlow, H. F., and M. Harlow. 1966. Learning to love. *American Scientist* 54 (3), 244–272.

Harlow, H., M. Harlow, and S. Suomi. 1971. From thought to therapy: lessons from a primate laboratory. *American Scientist* 59, 538–549.

Harris, Marvin, and Eric B. Ross, eds. 1987. *Food and Evolution: Toward a Theory of Human Food Habits*. Temple University Press, Philadelphia.

Harrison, G. A., J. M. Tanner, D. R. Pilbeam, and P. T. Baker. 1988. *Human Biology: An Introduction to Human Evolution, Variation, Growth, and Adaptability*. Oxford University Press, Oxford.

Harrison, R. J., and W. Montagna. 1973. *Man*. Appleton-Century-Crofts, New York.

Harvey, R. 1998. Cosmetic surgery's ugly side. *The Toronto Star,* 13 June.

Hauber, A. R. 1980. The social psychology of driving behaviour and the traffic environment: research on aggressive behaviour in traffic. *International Review of Applied Psychology* 29, 461–474.

Health Encyclopedia. 2001. Grieg's disease. *http://www.lineone.net/health_ encyclopaedia/dict/pages/g/299.htm*.

Hebb, D. O. 1949. Temperament in chimpanzees I: method of analysis. *Journal of Comparative and Physiological Psychology* 42, 192–206.

Herzog, D. B., and P. M. Copeland. 1985. Eating disorders. *New England Journal of Medicine* 313 (5), 295–303.

Hickman, Martin. 2001. McDonald's expresses "regret" in meaty fries row. *Reuters Business Report,* 24 May.

Hippel, Arndt von. 1994. *Human Evolutionary Biology: Human Anatomy and Physiology from an Evolutionary Perspective*. Stone Age Press, Anchorage.

Hirschfeld, R. M., M. B. Keller, S. Panico, B. S. Arons, D. Barlow, F. Davidoff, J. Endicott, J. Froom, M. Goldstein, J. M. Gorman, R. G. Marek, T. A. Maurer, R. Meyer, K. Phillips, J. Ross, T. L. Schwenk, S. S. Sharfstein, M. E. Thase, and R. J. Wyatt. 1997. The national depressive and manic-depressive association consensus statement on the undertreatment of depression. *Journal of the American Medical Association* 277 (4), 333–340.

Honig, A. S. 1983. Research in review: television and young children. *Young Child* 38, 63–74.

Hoyos, C. G. 1988. Mental load and risk in traffic behaviour. *Ergonomics* 31 (4), 571–584.

Hrdy, Sarah Blaffer. 1999. *Mother Nature: A History of Mothers, Infants, and Natural Selection*. Pantheon Books, New York.

Huffington, Arianna. 1999. Barking back at Prozac. *The Washington Times,* 20 January, final edition, section A, p. 15.

Hukshorn, C. J., W. H. Saris, M. S. Westerterp-Plantenga, A. R. Farid, F. J. Smith, and L. A. Campfield. 2000. Weekly subcutaneous pegylated recombinant native human leptin (peg-ob) administration in obese men. *Journal of Clinical Endocrinology & Metabolism* 85 (11), 4003–4009.

Hyman, S. E. 2000. The genetics of mental illness: implications for practice. *Bulletin of the World Health Organization* 78 (4), 455–463.

Insel, T. R., and J. T. Winslow. 1998. Serotonin and neuropeptides in affiliative behaviors. *Biological Psychiatry* 44 (3), 207–219.

Jackson, L. A. 1992. *Physical Appearance and Gender: Sociobiological and Sociocultural Perspectives*. State University of New York Press, Albany.

Jacobson, Michael F. 1999. *Liquid Candy: How Soft Drinks Are Harming Americans' Health*. Center for Science in the Public Interest, Washington, D.C.

James, Leon, and Diane Nahl. 2000. *Road Rage and Aggressive Driving: Steering Clear of Highway Warfare*. Prometheus Books, Amherst, N.Y.

Jequier, E., and L. Tappy. 1999. Regulation of body weight in humans. *Physiological Reviews* 79 (2), 451–480.

Joint, Matthew. 1995. *Road Rage*. The Automobile Association Group Public Policy Road Safety Unit, London.

Jones, John Morgan. 1997. *Physicians' Desk Reference to Pharmaceutical Specialties and Biologicals: Five Sections*. Medical Economics, Rutherford, N.J.

Judge, Debra S. 1995. American legacies and the variable life histories of women and men. *Human Nature* 6 (4), 291–323.

Kamara, Kalatu, Robert Eskay, and Thomas Castonguay. 1998. High-fat diets and stress responsivity. *Physiology & Behavior* 64 (1), 1–6.

Kandel, Eric R., James H. Schwartz, and Thomas M. Jessell, eds. 1991. *Principles of Neural Science*. Appleton & Lange, Norwalk, Conn.

Kaplan, J. R., M. F. Muldoon, S. B. Manuck, and J. J. Mann. 1997. Assessing the observed relationship between low cholesterol and violence-related mortality: implications for suicide risk. *Annals of the New York Academy of Sciences* 836, 57–80.

Kapur, S., T. Mieczkowski, and J. J. Mann. 1992. Antidepressant medications and the relative risk of suicide attempt and suicide. *Journal of the American Medical Association* 268 (24), 3441–3445.

Katz, M. M., A. Marsella, K. C. Dube, M. Olatawura, R. Takahashi, Y. Nakane, L. C. Wynne, T. Gift, J. Brennan, and N. Sartorius. 1988. On the expression of psychosis in different cultures: schizophrenia in an Indian and in a Nigerian community. *Culture, Medicine & Psychiatry* 12 (3), 331–355.

Kawanaka, Kenji. 1982. Further studies on predation by chimpanzees of the Mahale Mountains. *Primates* 23 (3), 364–384.

Kendler, K. S., and C. A. Prescott. 1999. A population-based twin study of lifetime major depression in men and women. *Archives of General Psychiatry* 56 (1), 39–44.

Kilbourne, Jean. 1994. Still killing us softly: advertising and the obsession with thinness. In Patricia Fallon and Melanie A. Katzman, eds., *Feminist Perspectives on Eating Disorders*. Guilford Press, New York, pp. 395–418.

King, Michael, and Marti Friedlander. 1972. *Moko: Maori Tattooing in the 20th Century*. Alister Taylor, Wellington, N.Z.

Kline, Wayne E. 1993. Bar Barbie. In Lucinda Ebersole and Richard Peabody, eds., *Mondo Barbie*. St. Martin's, New York, pp. 149–150.

Knapp, Bettina L. 1997. Beckett's *That Time*: Exile and "that double-headed monster . . . time." *Journal of Melanie Klein & Object Relations* 15 (3), 493–511.

Konner, Melvin. 1982. *The Tangled Wing: Biological Constraints on the Human Spirit*. Holt Rinehart and Winston, New York.

Koren, G., I. Nulman, and A. Addis. 1998. Outcome of children exposed in utero to fluoxetine: a critical review. *Depression and Anxiety* 8 (supplement 1), 27–31.

Kotchen, T. A., and R. M. Krauss. 1996. Dietary sodium and blood pressure. *Journal of the American Medical Association* 276 (18), 1468; discussion, 1469–1470.

Krantz, Grover S. 1968. Brain size and hunting ability in earliest man. *Current Anthropology* 9 (5), 450–451.

Kuhn, R. 1989. The discovery of modern antidepressants. *Psychiatric Journal of the University of Ottawa* 14 (1), 249–252.

Kumanyika, S. K. 1993. Special issues regarding obesity in minority populations. *Annals of Internal Medicine* 119 (7 part 2), 650–654.

Kunzle, David. 1982. *Fashion and Fetishism: A Social History of the Corset, Tight-Lacing, and Other Forms of Body-Sculpture in the West*. Rowman and Littlefield, Totowa, N.J.

Kusinitz, Marc. 1987. *Drugs & the Arts*. Chelsea House Publishers, New York.

Lajunen, Timo, Dianne Parker, and Heikki Summala. 1999. Does traffic congestion increase driver aggression? *Transportation Research Part F: Traffic Psychology & Behaviour* 2 (4), 225–236.

Landler, Mark, Walecia Konrad, Zachary Schiller, and Lois Therrien. 1991. What happened to advertising? *Business Week*, 23 September, #3232, 66–72.

Lane, R. 1993. Managing depression in general practice. *British Medical Journal* 306 (6872), 272–273.

Langlois, Judith H., and Lori A. Roggman. 1990. Attractive faces are only average. *Psychological Science* 1 (2), 115–121.

Langlois, Judith H., Jean M. Ritter, Lori A. Roggman, and Lesley S. Vaughn. 1991. Facial diversity and infant preferences for attractive faces. *Developmental Psychology* 27 (1), 79–84.

Langlois, Judith H., Lori A. Roggman, Rita J. Casey, Jean M. Ritter, et al. 1987. Infant preferences for attractive faces: rudiments of a stereotype? *Developmental Psychology* 23 (3), 363–369.

Langlois, Judith H., Lori A. Roggman, and Lisa Musselman. 1994. What is average and what is not average about attractive faces? *Psychological Science* 5 (4), 214–220.

Langlois, Judith H., Lori A. Roggman, and Loretta A. Rieser-Danner. 1990. Infants' differential social responses to attractive and unattractive faces. *Developmental Psychology* 26 (1), 153–159.

Larson, John. 1999. *Road Rage to Road-Wise: A Simple Step- by-Step Program to Help You Understand and Curb Road Rage in Yourself and Others*. Forge, New York.

Lee, I. M., C. C. Hsieh, and R. S. Paffenbarger, Jr. 1995. Exercise intensity and longevity in men: the Harvard alumni health study. *Journal of the American Medical Association* 273 (15), 1179–1184.

Leppa, C. J. 1990. Cosmetic surgery and the motivation for health and beauty. *Nursing Forum* 25 (1), 25–31.

Leroi-Gourhan, Arlette. 1975. The flowers found with Shanidar IV, a Neanderthal burial in Iraq. *Science* 190 (4214), 562–564.

Lex Service. 1997. Drugs, drink, and bad manners continue to plague our roads. Lex Service, London.

Liggett, John. 1974. *The Human Face*. Constable, London.

Livingstone, David. 1872. *Livingstone's Africa: Perilous Adventures and Extensive Discoveries in the Interior of Africa*. Hubbard Brothers, Philadelphia.

Livingstone, David, and Charles Livingstone. 1893. *Narrative of an Expedition to the Zambesi and Its Tributaries; and of the Discovery of the Lakes Shirwa and Nyassa, 1858–1864*. Harper & Brothers, New York.

Lofy, Mary Mead. 2000. A matter of time: power, control, and meaning in people's everyday experience of time. *Dissertation Abstracts International, A* (Humanities and Social Sciences) 60 (12–A), 4628.

Lonnqvist, F., L. Nordfors, and M. Schalling. 1999. Leptin and its potential role in human obesity. *Journal of Internal Medicine* 245 (6), 643–652.

Lorenz, Konrad, and Agnes von Cranach. 1996. *The Natural Science of the Human Species: An Introduction to Comparative Behavioral Research: The Russian Manuscript (1944–1948).* MIT Press, Cambridge, Mass.

Louis Mizell Inc. 1997. *Aggressive Driving,* AAA Foundation for Traffic Safety, Washington, D.C.

Lower, T., A. Girgis, and R. Sanson-Fisher. 1998. The prevalence and predictors of solar protection use among adolescents. *Preventive Medicine* 27 (3), 391–399.

Lupton, Cookie. 1993. Barbie meets the scariest fatso yet. In Lucinda Ebersole and Richard Peabody, eds., *Mondo Barbie.* St. Martin's, New York, p. 72.

MacGregor, G. A., and F. P. Cappuccio. 1996. Dietary sodium and blood pressure. *Journal of the American Medical Association* 276 (18), 1468–1469; discussion, 1469–1470.

MacGregor, Graham A., and Peter S. Sever. 1996. Salt—overwhelming evidence but still no action: can a consensus be reached with the food industry? *British Medical Journal* 312 (7041), 1287–1289.

Macmillan, John. 1975. *Deviant Drivers.* Saxon House/Lexington Books, Lexington, Mass.

Maddox, G. L., K. W. Back, and W. R. Liederman. 1968. Overweight as social deviance and disability. *Journal of Health & Social Behavior* 9 (4), 287–298.

Maddox, G. L., and V. Liederman. 1969. Overweight as a social disability with medical implications. *Journal of Medical Education* 44 (3), 214–220.

Magro, A. M. 1997. Why Barbie is perceived as beautiful. *Perceptual & Motor Skills* 85 (1), 363–374.

Majeroni, B. A., and A. Hess. 1998. The pharmacologic treatment of depression. *Journal of the American Board of Family Practice* 11 (2), 127–139.

Malamba, S. S., H. U. Wagner, G. Maude, M. Okongo, A. J. Nunn, J. F. Kengeya-Kayondo, and D. W. Mulder. 1994. Risk factors for HIV-1 infection in adults in a rural Ugandan community: a case-control study. *AIDS* 8 (2), 253–257.

Malinowski, Bronislaw. 1929. Practical anthropology. *Africa: Rivista Trimestrale di Studi e Documentazione dell'Istituto Italo-Africano* 2, 22–38.

———. 1961. *Argonauts of the Western Pacific: An Account of Native Enterprise and Adventure in the Archipelagoes of Melanesian New Guinea.* Dutton, New York.

Manda, Valerie. 2001, Sabin and OPV. Personal communication, 7 March.

Manolis, D. C. 1995. The perils of Prozac. *Minnesota Medicine* 78 (1), 19–23.

Many bicyclists came to grief. 1896. *New York Times,* 31 May, p. 2.

Marano, Hara Estroff. 1999. Depression: beyond serotonin. *Psychology Today* 31 (April), 30–36, 72.

Marieb, E. N. 1995. *Human Anatomy and Physiology.* Benjamin/Cummings, Redwood City, Calif.

Marquardt, Carl. 1984. *The Tattooing of Both Sexes in Samoa*. R. McMillan, Papakura.

Marsella, A. J. 1988. Cross-cultural research on severe mental disorders: issues and findings. *Acta Psychiatrica Scandinavica*, supplementum, 344, 7–22.

Marsh, P., and P. Collett. 1987. The car as a weapon. *ETC: A Review of General Semantics* 44 (2), 146–151.

Martinez, Ricardo. 1997. Statement of the Honorable Ricardo Martinez, M.D., administrator, National Highway Traffic Safety Administration. U.S. House of Representatives, Subcommittee on Surface Transportation, Washington, D.C.

Mayr, Ernst. 1988. *Toward a New Philosophy of Biology: Observations of an Evolutionist*. Harvard University Press, Cambridge, Mass.

———. 1997. *This Is Biology: The Science of the Living World*. Harvard University Press, Cambridge, Mass.

Mazur, A. 1986. U.S. trends in feminine beauty and overadaptation. *Journal of Sex Research* 22 (3), 281–303.

McAleer, Kevin. 1994. *Dueling: The Cult of Honor in Fin-de-Siècle Germany*. Princeton University Press, Princeton, N.J.

McCance, R. A. 1936. Medical problems in mineral metabolism III: experimental human salt deficiency. *Lancet* 230, 823–830.

———. 1938. The effect of salt deficiency in man on the volume of the extracellular fluid and on the composition of sweat, saliva, gastric juice and cerebrospinal fluid. *Journal of Physiology* 92, 208–218.

McCurdy, John A. 1992. *The Complete Guide to Cosmetic Facial Surgery*. Lifetime Books, Hollywood, Fla.

McKim, Jenifer. 2001. Now at the mall: "parking lot rage"; short-tempered drivers vying for a space end up fighting or defacing cars. *Washington Post*, 20 December, final edition, section A, p. 19.

McKinney, W. T., S. J. Suomi, and H. Harlow. 1971. The sad ones. *Psychology Today* (May), 61–68.

McMurray, Teresa M., and Charles T. Snowdon. 1977. Sodium preferences and responses to sodium deficiency in rhesus monkeys. *Physiological Psychology* 5 (4), 477–482.

McNeal, James U. 1998. Tapping the three kids' markets. *American Demographics* 20 (4), 36–42.

———. 1999. *The Kids Market: Myths and Realities*. Paramount Market, Ithaca, N.Y.

Mead, Margaret. 1928. *Coming of Age in Samoa: A Psychological Study of Primitive Youth for Western Civilisation*. W. Morrow, New York.

Merker, M. 1904. *Die Masai: Ethnographische Monographie Eines Ostafrikanischen Semitenvolkes*. D. Reimer, Berlin.

Messenger, J. C. 1959. Religious acculturation among the Anang Ibibio. In W. Bascom and M. Herskovitz, eds., *Continuity and Change in African Cultures*. University of Chicago Press, Chicago, pp. 279–299.

Messerli, F. H., and R. E. Schmieder. 1996. Dietary sodium and blood pressure.

Journal of the American Medical Association 276 (18), 1469; discussion, 1469–1470.

Mhanna, M. J., J. B. Bennet, 2nd, and S. D. Izatt. 1997. Potential fluoxetine chloride (Prozac) toxicity in a newborn. *Pediatrics* 100 (1), 158–159.

Minino, Arialdi M., and Betty L. Smith. 2001. Deaths: Preliminary data for 2000. *National Vital Statistics Reports* 49 (12), 14–16, table 2. National Center for Health Statistics, Hyattsville, Md.

Mokdad, A. H., M. K. Serdula, W. H. Dietz, B. A. Bowman, J. S. Marks, and J. P. Koplan. 1999. The spread of the obesity epidemic in the United States, 1991–1998. *Journal of the American Medical Association* 282 (16), 1519–1522.

Montgomery, S. A. 1997. Suicide and antidepressants. *Annals of the New York Academy of Sciences* 836, 329–338.

Mooar, Brian, and Fern Shen. 1998. 2nd driver files complaint saying Tyson assaulted him. *Washington Post,* 2 September, final edition, metro section B, p. 4.

Moore, G. E. 1922. *Principia Ethica.* Cambridge University Press, Cambridge.

Moran, Greg. 1984. Vigilance behaviour and alarm calls in a captive group of meerkats, *Suricata suricatta. Zeitschrift für Tierpsychologie* 65 (3), 228–240.

Morris, Desmond. 1977. *Manwatching: A Field Guide to Human Behaviour.* Cape, London.

Morris, Desmond, Peter Collett, Peter Marsh, and Marie O'Shaughnessy. 1979. *Gestures.* Stein and Day, New York.

Morrison, D. E., and C. P. Holden. 1971. The burning bra: the American breast fetish and women's liberation. In P. K. Manning, ed., *Deviance and Change.* Prentice-Hall, Englewood Cliffs, N.J.

Muldoon, M. F., J. R. Kaplan, S. B. Manuck, and J. J. Mann. 1992. Effects of a low-fat diet on brain serotonergic responsivity in cynomolgus monkeys. *Biological Psychiatry* 31 (7), 739–742.

Nagatsuka, Y. 1989. The current situation of traffic psychology in Japan. *Applied Psychology: An International Review* 38 (4), 423–442.

Najman, J. M., D. Klein, and C. Munro. 1982. Patient characteristics negatively stereotyped by doctors. *Social Science & Medicine* 16 (20), 1781–1789.

National Center for Health Statistics (U.S.). 1996. *The Third National Health and Nutrition Examination Survey (NHANES III, 1988–94). Reference Manuals and Reports.* U.S. Dept. of Health and Human Services, Centers for Disease Control and Prevention, National Center for Health Statistics, U.S. Government Printing Office, Washington, D.C.

———. 2001. *National Ambulatory Medical Care Survey, 1999: Summary.* U.S. Public Health Service, Centers for Disease Control and Prevention, Washington, D.C.

National Highway Traffic Safety Administration and Federal Highway Administration. 1999. *Summary Report: Aggressive Driving and the Law.* Washington, D.C.

National Traffic Safety Board. 1982. *Air Florida, Inc., Boeing 737-222, N62AF, Collision with 14th Street Bridge, near Washington National Airport, Washington, D.C., 13 January 1982, AAR-82-08.*

Neergaard, Lauran. 1999. FDA approves anti-depressant for dogs. *The Columbian*, 5 January, world/national news.

Nerenberg, Arnold P. 2001. America's leading authority on "road rage" and aggressive driving. *http://www.drnerenberg.com/roadrage.htm*.

Nesse, Randolph M., and George C. Williams. 1995. *Why We Get Sick: The New Science of Darwinian Medicine*. Times Books, New York.

Newman, W. P., and R. G. Brodows. 1983. Insulin action during acute starvation: evidence for selective insulin resistance in normal man. *Metabolism: Clinical & Experimental* 32 (6), 590–596.

Nicholas, Anne. 1994. *The Art of the New Zealand Tattoo*. Carol Pub. Group, Secaucus, N.J.

Nissenbaum, Stephen, and Sylvester Graham. 1980. *Sex, Diet, and Debility in Jacksonian America: Sylvester Graham and Health Reform*. Greenwood Press, Westport, Conn.

Nowak, Ronald M., and Ernest P. Walker. 1991. *Walker's Mammals of the World*. Vol. 1. Johns Hopkins University Press, Baltimore.

Object to the abuse of automobiling. 1902. *Motor World* 4, 662.

Office of Management and Budget. 2001. *Budget of the United States Government, 2001*. Washington, D.C.

Ogletree, S. M., S. W. Williams, P. Raffeld, B. Mason, and K. Fricke. 1990. Female attractiveness and eating disorders: do children's television commercials play a role? *Sex Roles* 22, 791–797.

Olfson, M., and G. L. Klerman. 1992. The treatment of depression: prescribing practices of primary care physicians and psychiatrists. *Journal of Family Practice* 35 (6), 627–635.

———. 1993. Trends in the prescription of antidepressants by office-based psychiatrists. *American Journal of Psychiatry* 150 (4), 571–577.

Olfson, M., S. C. Marcus, H. A. Pincus, J. M. Zito, J. W. Thompson, and D. A. Zarin. 1998. Antidepressant prescribing practices of outpatient psychiatrists. *Archives of General Psychiatry* 55 (4), 310–316.

Olsen, Nora. 2000. Varietal differences bring fry color challenge. *Potato Grower* 29 (12).

Orbach, Susie. 1982. *Fat Is a Feminist Issue II: A Program to Conquer Compulsive Eating*. Berkley Books, New York.

———. 1986. *Hunger Strike: An Anorectic's Struggle as a Metaphor for Our Age*. Norton, New York.

Owen, M. J., and M. J. Mullan. 1990. Molecular genetic studies of manic-depression and schizophrenia. *Trends in Neurosciences* 13 (1), 29–31.

Oxford University Press. 1996. *The Oxford Modern English Dictionary*. Oxford University Press, New York.

Packer, C. 1977. Reciprocal altruism in *Papio anubis*. *Nature* 265, 441–443.

Papiernik, Richard L. 2001. Sales expected to hit $399 billion. *Nation's Restaurant News* 38 (1), 1–3.

Parker, F. D., and J. B. Welch. 1991. Field comparisons of attractants for the screwworm fly (Diptera, Calliphoridae) in a tropical dry forest in Costa Rica. *Journal of Economic Entomology* 84 (4), 1189–1195.

Parry, Meyer Hai. 1968. *Aggression on the Road: A Pilot Study of Behaviour in the Driving Situation*. Tavistock Publications, London.

Patton, W. S. 1922. Notes on the myiasis producing diptera of man and animals. *Bulletin of Entomological Research* 12, 239–261.

Paykel, E. S., A. Tylee, A. Wright, R. G. Priest, S. Rix, and D. Hart. 1997. The defeat depression campaign: psychiatry in the public arena. *American Journal of Psychiatry* 154 (6 supplement), 59–65.

Pear, Robert. 2001. Spending on prescription drugs increases by almost 19 percent. *New York Times,* 8 May, final edition, section A, p. 1.

Peiss, Kathy Lee. 1998. *Hope in a Jar: The Making of America's Beauty Culture*. Metropolitan Books, New York.

Pennell, Amanda E., and Kevin D. Browne. 1999. Film violence and young offenders. *Aggression & Violent Behavior* 4 (1), 13–28.

Peterson, G. W., and D. F. Peters. 1983. Adolescents' construction of social reality—the impact of television and peers. *Youth and Society* 15, 67–85.

Peterson, R. T. 1987. Bulimia and anorexia in an advertising context. *Journal of Business Ethics* 6, 495–504.

Physicians Committee for Responsible Medicine. 2000. Doctors announce final victory in dietary guidelines lawsuit. Washington, D.C.

Pi-Sunyer, F. X. 1993. Medical hazards of obesity. *Annals of Internal Medicine* 119 (7 Pt 2), 655–660.

———. 1994. The fattening of America. *Journal of the American Medical Association* 272 (3), 238–239.

Platt, S., ed. 1992. *Respectfully Quoted: A Dictionary of Quotations*. Barnes & Noble, New York.

Pond, Caroline M. 1998. *The Fats of Life*. Cambridge University Press, Cambridge.

Potato Association of America. 2001. Potato varieties. *http://www.css.orst.edu/classes/CSS322/potato.htm.*

Prescott, H. M. 1993. Anorexia nervosa. In K. F. Kiple, ed., *The Cambridge World History of Human Disease*. Cambridge University Press, New York, pp. 577–582.

Price, J. H., S. M. Desmond, R. A. Krol, F. F. Snyder, and J. K. O'Connell. 1987. Family practice physicians' beliefs, attitudes, and practices regarding obesity. *American Journal of Preventive Medicine* 3 (6), 339–345.

Prielipp, Richard C., and Robert C. Morell. 1999. Liposuction in the United States: beauty and the beast. *Anesthesia Patient Safety Foundation Newsletter* 14 (2), 18–19.

Randle, H. W. 1997. Suntanning: differences in perceptions throughout history. *Mayo Clinic Proceedings* 72 (5), 461–466.

Rao, R. B., S. F. Ely, and R. S. Hoffman. 1999. Deaths related to liposuction. *New England Journal of Medicine* 340 (19), 1471–1475.

Rasa, O. Anne. 1976. Invalid care in the dwarf mongoose (*Helogale undulata rufula*). *Zeitschrift für Tierpsychologie* 42 (4), 337–342.

———. 1983. A case of invalid care in wild dwarf mongooses. *Zeitschrift für Tierpsychologie* 62 (3), 235–240.

Rathbone, D. B., and J. C. Huckabee. 1999. *Controlling Road Rage: A Literature Review and Pilot Study*. AAA Foundation for Traffic Safety, Washington, D.C.

Ravussin, E., M. E. Valencia, J. Esparza, P. H. Bennett, and L. O. Schulz. 1994. Effects of a traditional lifestyle on obesity in Pima Indians. *Diabetes Care* 17 (9), 1067–1074.

Reuters Business Report. 2001. McDonald's sued by another group over beefy fries. *Reuters Business Report*, 12 June.

Reynolds, K. D., J. M. Blaum, P. M. Jester, H. Weiss, S. J. Soong, and R. J. Diclemente. 1996. Predictors of sun exposure in adolescents in a southeastern U.S. population. *Journal of Adolescent Health* 19 (6), 409–415.

Richardson, J., and A. L. Kroeber. 1940. Three centuries of women's dress fashions: a quantitative analysis. *Anthropological Records* 5, 111–150.

Richins, M. L. 1981. Social comparison and idealized images of advertising. *Journal of Consumer Research* 18, 71–83.

Ridley, M. 1996. *Evolution*. Blackwell Science, Cambridge.

Rischer, Carl E., and Thomas A. Easton. 1992. *Focus on Human Biology*. Harper-Collins, New York.

Rogers, B. O. 1971. A brief history of cosmetic surgery. *Surgical Clinics of North America* 51 (2), 265–288.

———. 1971. A chronologic history of cosmetic surgery. *Bulletin of the New York Academy of Medicine* 47 (3), 265–302.

Rotella, Sebastian, and Chris Kraul. 1995. Tank's driver beset by drugs, money problems. *Los Angeles Times*, 19 May, home edition, part A, p. 1.

Rothenberg, M. D. 1983. The role of television in shaping the attitudes of children. *Journal of the American Academy of Child Psychology* 22, 86–87.

Rothschild, A. J. 1988. Biology of depression. *Medical Clinics of North America* 72 (4), 765–790.

Rubenstein, Adam J., Lisa Kalakanis, and Judith H. Langlois. 1999. Infant preferences for attractive faces: a cognitive explanation. *Developmental Psychology* 35 (3), 848–855.

Rushton, J. L., S. J. Clark, and G. L. Freed. 2000. Pediatrician and family physician prescription of selective serotonin reuptake inhibitors. *Pediatrics* 105 (6), E82.

Rushton, J. L., and J. T. Whitmire. 2001. Pediatric stimulant and selective serotonin reuptake inhibitor prescription trends: 1992 to 1998. *Archives of Pediatrics & Adolescent Medicine* 155 (5), 560–565.

Ryan, N. D. 1992. The pharmacologic treatment of child and adolescent depression. *Psychiatric Clinics of North America* 15 (1), 29–40.

Sack, Robert David. 1986. *Human Territoriality: Its Theory and History*. Cambridge University Press, Cambridge.

Sagan, Carl. 1977. *The Dragons of Eden: Speculations on the Evolution of Human Intelligence*. Random House, New York.

Samuels, Curtis A., George Butterworth, Tony Roberts, Lida Graupner, et al. 1994. Facial aesthetics: babies prefer attractiveness to symmetry. *Perception* 23 (7), 823–831.

Sanders, Clinton R. 1989. *Customizing the Body: The Art and Culture of Tattooing*. Temple University Press, Philadelphia.

Sapolsky, Robert M. 1994. *Why Zebras Don't Get Ulcers: A Guide to Stress, Stress Related Diseases, and Coping*. W. H. Freeman, New York.

SBRnet. 2001. Total market consumer expenditures. *http://www.sbrnet.com/Research/Research.cfm?subRID=126*.

Schappert, S. M., and C. Nelson. 1999. National ambulatory medical care survey, 1995–96: summary. *Vital Health Statistics* 13 (142), i–vi, 1–122.

Schlosser, Eric. 2001. *Fast Food Nation: The Dark Side of the All-American Meal*. Houghton Mifflin, Boston.

Schrank, David, Tim Lomax, and Bernie Fette. 1999. *The 1999 Annual Mobility Report*. Texas A&M University, College Station, Tex.

Schroeder, M. 1991. The diet business is getting a lot skinnier. *Business Week*, 24 June, #3219, pp. 132–134.

Schulkin, J. 1986. The evolution and expression of salt appetite. In G. de Caro, A. N. Epstein, and M. Massi, eds., *The Physiology of Thirst and Appetite*. Plenum, New York.

———. 1991. *Sodium Hunger: The Search for a Salty Taste*. Cambridge University Press, Cambridge.

Schulkin, Jay, et al. 1984. Salt hunger in the rhesus monkey. *Behavioral Neuroscience* 98 (4), 753–756.

Schwartz, Hillel. 1986. *Never Satisified: A Cultural History of Diets, Fantasies, and Fat*. Free Press, New York.

Scotti, Ciro. 1999. Just what American needs: Major League Baseball Barbie. *Business Week*, 9 June. *http://www.businessweek.com/bwdaily/dnflash/june1999/nf90609d.htm*.

Seibold, I., and A. J. Helbig. 1995. Evolutionary history of New and Old World vultures inferred from nucleotide sequences of the mitochondrial cytochrome b gene. *Philosophical Transactions of the Royal Society of London, series B—Biological Sciences* 350 (1332), 163–178.

Senut, B., M. Pickford, D. Gommery, P. Mein, C. Cheboi, and Y. Coppens. 2001. First hominid from the Miocene (Lukeino Formation, Kenya). *Comptes Rendus de L'académie des Sciences, Paris* 332, 137–144.

Shackelford, Todd K., and David M. Buss. 1997. Anticipation of marital dissolution as a consequence of spousal infidelity. *Journal of Social & Personal Relationships* 14 (6), 793–808.

Shakespeare, William, and George MacDonald. 1885. *The Tragedie of Hamlet, Prince of Denmarke; a Study with the Text of the Folio of 1623*. Longmans Green and Co., London.

Shorter, Edward. 1997. *A History of Psychiatry: From the Era of the Asylum to the Age of Prozac*. John Wiley & Sons, New York.

Sibley, C. G., and J. E. Ahlquist. 1987. DNA hybridization evidence of hominoid phylogeny: results from an expanded data set. *Journal of Molecular Evolution* 26 (1–2), 99–121.

Sigerist, Henry E. 1987. *A History of Medicine: Early Greek, Hindu, and Persian Medicine*. Oxford University Press, New York.

Simmons, D. R. 1986. *Ta Moko: The Art of Maori Tattoo*. Reed Methuen, Auckland, N.Z.

Simon, G. E., M. VonKorff, T. B. Ustun, R. Gater, O. Gureje, and N. Sartorius. 1995. Is the lifetime risk of depression actually increasing? *Journal of Clinical Epidemiology* 48 (9), 1109–1118.

Simonton, D. K. 1977. Women's fashions and war: a quantitative comment. *Social Behavior & Personality* 5 (2), 285–288.

Skinner, B. F. 1948. *Walden Two*. Macmillan, New York.

Smith, Adam. 1933. *The Wealth of Nations*. J. M. Dent & Sons, London.

Smith, E. O. 1987. Deception and evolutionary biology. *Cultural Anthropology* 2, 50–64.

———. 1999. High heels and evolution. *Psychology, Evolution, and Gender* 1 (3), 245–277.

Smith, E. O., and W. S. Helms. 1999. Natural selection and high heels. *Foot and Ankle International* 20 (1), 55–57.

Smith, J. R. 1969. Television violence and driving behavior. *Educational Broadcasting Review* 3, 23–28.

Smith, John, and Philip L. Barbour. 1986. *The Complete Works of Captain John Smith (1580–1631)*. University of North Carolina Press, Chapel Hill.

Smith, Michael F. 1994. *Research Agenda for an Improved Novice Driver Education Program: Report to Congress*. U.S. Department of Transportation, Washington, D.C.

Solecki, R. S. 1971. *Shanidar: The First Flower People*. Knopf, New York.

Sorensen, J., and P. Kind. 1995. Modelling cost-effectiveness issues in the treatment of clinical depression. *IMA Journal of Mathematics Applied in Medicine & Biology* 12 (3–4), 369–385.

Spennemann, Dirk R. V. 1992. *Marshallese Tattoos*. Republic of the Marshall Islands Historic Preservation Office, Majuro Atoll.

Stamler, J. 1996. Dietary sodium and blood pressure. *Journal of the American Medical Association* 276 (18), 1467; discussion, 1469–1470.

Stanford, C. B., J. Wallis, H. Matama, and J. Goodall. 1994. Patterns of predation by chimpanzees on red colobus monkeys in Gombe National Park, 1982–1991. *American Journal of Physical Anthropology* 94, 213–228.

Stark, Phyllis. 1994. A history of radio broadcasting. *Billboard* 107 (1 November), 121–125.

Steinhausen, H. C., C. Rauss-Mason, and R. Seidel. 1991. Follow-up studies of anorexia nervosa: a review of four decades of outcome research. *Psychological Medicine* 21 (2), 447–454.

Stephens, D. L., R. P. Hill, and C. Hanson. 1994. The beauty myth and female consumers: the controversial role of advertising. *Journal of Consumer Affairs* 28, 137–153.

Stephenson, G. R. 1967. Cultural acquisition of a specific learned response among rhesus monkeys. In D. Starck, R. Schneider, and H. J. Kuhn, eds., *Progress in Primatology*. G. Fischer, Stuttgart, pp. 279–288.

Stevens, Anthony, and John Price. 1996. *Evolutionary Psychiatry: A New Beginning*. Routledge, London.

Subraman, Belinda. 1993. The black lace panties triangle. In Lucinda Ebersole and Richard Peabody, eds., *Mondo Barbie*. St. Martin's, New York, pp. 82–83.

Sumner, William Graham. 1959. *Folkways: A Study of the Sociological Importance of Usages, Manners, Customs, Mores, and Morals*. Dover Publications, New York.

Symons, D. 1979. *The Evolution of Human Sexuality*. Oxford University Press, New York.

———. 1995. Beauty is in the adaptations of the beholder: the evolutionary psychology of female attractiveness. In P. R. Abramson and S. D. Pinkerton, eds., *Sexual Nature, Sexual Culture*. University of Chicago Press, Chicago, pp. 80–118.

Szegedy-Maszak, Marianne. 2001. The career of a celebrity pill. *U.S. News & World Report* 131 (5), 38.

Tabassam, Waheeda. 1994. A study of causal factors of antisocial behaviour in school children. *Bangladesh Journal of Psychology* 14, 69–75.

Tan, A. S. 1979. TV beauty ads and role expectations of adolescent female viewers. *Journalism Quarterly* 56, 283–288.

Taylor, Rod. 1997. The Beanie factor. *Brandweek* 38 (24), 22–27.

Teleki, G. 1973. *The Predatory Behavior of Wild Chimpanzees*. Bucknell University Press, Lewisburg, Pa.

Teller, Esther R. 2000. Hairdressing. *http://www.comptons.com/ceo99-cgi/article?'fastweb?getdoc+viewcomptons+A+3537+0++'history%20of%20hair'*

Thakore, J. H., and D. N. John. 1996. Prescription of antidepressants by general practitioners: recommendations by FHSAS and health boards. *British Journal of General Practice* 46 (407), 363–364.

Thelle, D. S. 1996. Salt and blood pressure revisited. *British Medical Journal* 312 (7041), 1240–1241.

Thomas, D. B., and R. L. Mangan. 1989. Oviposition and wound-visiting behaviour of the screwworm fly, *Cochliomyia hominivorax* (Diptera: Calliphoridae). *Annals of the Entomological Society of America* 82, 526–534.

Thomas, P. R., and Institute of Medicine (U.S.), Committee to Develop Criteria for Evaluating the Outcomes of Approaches to Prevent and Treat Obesity. 1995. *Weighing the Options: Criteria for Evaluating Weight-Management Programs*. National Academy Press, Washington, D.C.

Thompson, C., R. C. Peveler, D. Stephenson, and J. McKendrick. 2000. Compliance with antidepressant medication in the treatment of major depressive disorder in primary care: a randomized comparison of fluoxetine and a tricyclic antidepressant. *American Journal of Psychiatry* 157 (3), 338–343.

Thornhill, Randy, and Karl Grammer. 1999. The body and face of woman:

one ornament that signals quality? *Evolution & Human Behavior* 20 (2), 105–120.

Tinbergen, N. 1963. On aims and methods of ethology. *Zeitschrift für Psychologie* 20, 410–433.

Tinbergen, Niko, and Time-Life Books. 1978. *Animal Behavior.* Time-Life Books, Alexandria, Va.

Townsend, J. M., and G. D. Levy. 1990. Effects of potential partners' physical attractiveness and socioeconomic status on sexuality and partner selection. *Archives of Sexual Behavior* 19, 149–164.

Trayhurn, P., N. Hoggard, J. G. Mercer, and D. V. Rayner. 1999. Leptin: fundamental aspects. *International Journal of Obesity & Related Metabolic Disorders* 23 (supplement 1), 22–28.

Trinkaus, Erik. 1983. *The Shanidar Neandertals.* Academic Press, New York.

Trinkaus, Erik, and Pat Shipman. 1994. *The Neandertals: Of Skeletons, Scientists, and Scandal.* Vintage Books, New York.

Trivers, Robert L. 1971. The evolution of reciprocal altruism. *The Quarterly Review of Biology* 46, 35–57.

———. 1972. Parental investment and sexual selection. In B. G. Campbell, ed., *Sexual Selection and the Descent of Man.* Aldine, Chicago, pp. 136–179.

Trivers, R. L., and D. E. Willard. 1973. Natural selection of parental ability to vary the sex ratio of offspring. *Science* 179, 90–92.

Trustees of Social Security and Medicare Boards. 2000. Status of the Social Security and Medicare programs: a summary of the 2000 annual reports. *http://www.ssa.gov/OACT/TRSUM/trsummary.html.*

Tylor, Edward Burnett. 1871. *Primitive Culture: Researches into the Development of Mythology, Philosophy, Religion, Art, and Custom.* J. Murray, London.

Uehara, S. 1986. Sex and group differences in feeding on animals by wild chimpanzees in the Mahale Mountains National Park, Tanzania. *Primates* 27, 1–13.

———. 1997. Predation on mammals by the chimpanzee (*Pan troglodytes*). *Primates* 38 (2), 193–214.

United Press International. 2001. McDonald's secret sauce secret no more. 13 August.

University of Alberta Health Centre. 2001. Health information page: food, weight and body image. *http://www.ualberta.ca/healthinfo.*

USDA. 2000. *Dietary guidelines for Americans.* Home and Garden Report 232. United States Department of Agriculture, Washington, D.C.

Van den Berghe, Pierre L. 1977. Territorial behavior in a natural human group. *Social Science Information* 16 (3 supplement 4), 419–430.

Vaughn, Brian E., and Judith H. Langlois. 1983. Physical attractiveness as a correlate of peer status and social competence in preschool children. *Developmental Psychology* 19 (4), 561–567.

Vest, Jason, Warren Cohen, Mike Tharp, Anna Mulrine, Mary Lord, Brendan I. Koerner, Barbra Murray, and Steven D. Kaye. 1997. Road rage. *U.S. News & World Report* 122, 24–32.

Wairere Iti, Tame, Pita Turei, and Nicole MacDonald. 1999. *Moko—Maori Tattoo*. Edition Stemmle, Zurich.

Weddle, Jeff. 1993. Hell's Angel Barbie. In Lucinda Ebersole and Richard Peabody, eds., *Mondo Barbie*. St. Martin's, New York, pp. 167–168.

Weigman, O., and van E.G.M. Schie. 1998. Video game playing and its relations with aggressive and prosocial behavior. *British Journal of Social Psychology* 37, 367–378.

Weiss, S. R., B. H. McFarland, G. A. Burkhart, and P. T. Ho. 1998. Cancer recurrences and secondary primary cancers after use of antihistamines or antidepressants. *Clinical Pharmacology & Therapeutics* 63 (5), 594–599.

Whelton, P. K., J. D. Cohen, and W. B. Applegate. 1996. Dietary sodium and blood pressure. *Journal of the American Medical Association* 276 (18), 1467–1468; discussion, 1469–1470.

Whitlock, Francis Antony. 1971. *Death on the Road: A Study in Social Violence*. Tavistock Publications, London.

Whitten, P. L. 2001. Personal communication, Atlanta, Ga.

Wilkes, M. S., B. H. Doblin, and M. F. Shapiro. 1992. Pharmaceutical advertisements in leading medical journals: experts' assessments. *Annals of Internal Medicine* 116 (11), 912–919.

Wilkinson, Gerald S. 1984. Reciprocal food sharing in the vampire bat. *Nature* 308 (5955), 181–184.

———. 1988. Reciprocal altruism in bats and other mammals. *Ethology & Sociobiology* 9 (2–4), 85–100.

Williams, T. Harry. 1989. *Huey Long*. Vintage, New York.

Williamson, D. F., M. K. Serdula, R. F. Anda, A. Levy, and T. Byers. 1992. Weight loss attempts in adults: goals, duration, and rate of weight loss. *American Journal of Public Health* 82 (9), 1251–1257.

Willis, David K. 1997. *Research on the Problem of Violent Aggressive Driving*. House of Representatives, Washington, D.C.

Wilson, D. R. 1998. Evolutionary epidemiology and manic depression. *British Journal of Medical Psychology* 71 (part 4), 375–395.

Wilson, E. 1985. *Adorned in Dreams: Fashion and Modernity*. Virago Press, London.

Wilson, E. O. 1975. *Sociobiology: The New Synthesis*. Harvard University Press, Cambridge, Mass.

Wink, M. 1995. Phylogeny of Old and New World vultures (Aves: Accipitridae and Cathartidae) inferred from nucleotide sequences of the mitochondrial cytochrome b gene. *Zeitschrift für Naturforschung C—Journal of Biosciences* 50 (11–12), 868–882.

Wisner, K. L., A. J. Gelenberg, H. Leonard, D. Zarin, and E. Frank. 1999. Pharmacologic treatment of depression during pregnancy. *Journal of the American Medical Association* 282 (13), 1264–1269.

Wissler, Clark, and Lucy Wales Kluckhohn. 1966. *Indians of the United States*. Doubleday, Garden City, N.Y.

Wittenberger, James F. 1981. *Animal Social Behavior*. Duxbury Press, Boston.

Wolf, A. M., and G. A. Colditz. 1998. Current estimates of the economic cost of obesity in the United States. *Obesity Research* 6 (2), 97–106.

Wolpoff, Milford H. 1999. *Paleoanthropology*. McGraw-Hill, New York.

Wrangham, R. W., and E. Van-Zinnicq-Bergmann-Riss. 1990. Rates of predation on mammals by Gombe chimpanzees, 1972–1975. *Primates* 31, 157–170.

Zuckerman, D. M., and B. S. Zuckerman. 1985. Television's impact on children. *Pediatrics* 75, 233–240.

Index

About the Author

E. O. Smith has edited three books, most recently *Evolutionary Medicine*. He teaches biological anthropology at Emory University in Atlanta. He lives in Atlanta with his wife, Cindy Gelb, and their three Welsh corgis, Woody, Ivy, and Farrah.